INVESTIGATING CHANGE IN AMERICAN SOCIETY

Exploring Social Trends with US Census Data and StudentChip

William H. Frey
University of Michigan

with

Cheryl L. First

Software Created by Zeta Data
Hanover, New Hampshire

Wadsworth Publishing Company
I(T)P® An International Thomson Publishing Company

Belmont, CA • Albany, NY • Bonn • Boston • Cincinnati • Detroit • Johannesburg • London • Madrid
Melbourne • Mexico City • New York • Paris • San Francisco • Singapore • Tokyo • Toronto • Washington

For more information, contact Wadsworth Publishing Company, 10 Davis Drive, Belmont, CA 94002, or electronically at http://www.thomson.com/wadsworth.html

International Thomson Publishing Europe
Berkshire House 168-173
High Holborn
London, WC1V 7AA, England

Thomas Nelson Australia
102 Dodds Street
South Melbourne 3205
Victoria, Australia

Nelson Canada
1120 Birchmount Road
Scarborough, Ontario
Canada M1K 5G4

International Thomson Publishing GmbH
Königswinterer Strasse 418
53227 Bonn, Germany

International Thomson Editores
Campos Eliseos 385, Piso 7
Col. Polanco
11560 México D.F. México

International Thomson Publishing Asia
221 Henderson Road
#05-10 Henderson Building
Singapore 0315

International Thomson Publishing Japan
Hirakawacho Kyowa Building, 3F
2-2-1 Hirakawacho
Chiyoda-ku, Tokyo 102, Japan

International Thomson Publishing Southern Africa
Building 18, Constantia Park
240 Old Pretoria Road
Halfway House, 1685 South Africa

ISBN 0-534-52344-7

CONTENTS

PREFACE

This book's approach is based on the premise that engaging, *hands on* data analysis cannot be introduced early enough in the social science curriculum. Or, to borrow the motto of a well-known athletic shoe company, "Just do it!" The investigations in this book were intentionally developed to be accessible, relevant, and user-friendly enough (that is, both student-friendly and teacher-friendly) for use in a wide range of substantive courses in sociology and related disciplines.

HOW TO USE THIS BOOK IN YOUR COURSE

The topics covered in this book can be integrated into a variety of courses including *Introductory Sociology, Social Problems, The Family, Social Stratification, Race and Ethnic Studies, Gender Studies, Demography, Research Methods,* and *American Society*. To incorporate these topics into your course, you can mix and match them so that they follow the logic of your syllabus. For example, a course on *The Family* might use Topic Five: Marriage, Divorce, Cohabitation, and Childbearing; Topic Seven: Households and Family; Topic Eight: Poverty; and Topic Nine: Children, while a course on *Social Stratification* course might use Topic Two: Race and Ethnic Inequality; Topic Three: Immigrant Assimilation; Topic Four: Labor Force; Topic Six: Gender Inequality; and Topic Eight: Poverty. You can choose the topics you will use according to the goals you have for your course, and you can present them in whatever order most successfully achieves these goals. Furthermore, the topics are flexible enough to be used with a variety of texts and/or readings on these topics.

The subject matter within each Topic follows a sequence consistent with the way it is typically covered in existing courses. For example, in both Topic Two (Race and Ethnic Inequality) and Topic Six (Gender Inequality), inequalities are first explored with respect to educational attainment, then occupation, then earnings — following the classic status attainment model which relates race or gender differences in earnings to differences in education and occupation. Students will develop the skills they need to test and prove theories by going through the exercises in this order.

At the end of each Investigation Topic there are *Think Tank* questions, which are less structured and can be the basis for class discussions or team-based reports. Please feel free to design your own questions with the datasets that are provided with this book. The possibilities are nearly unlimited! The last section of the book titled *Guide to Datasets* contains a complete list of all the datasets and their variables. Also, in the back of the book, is a list of useful World Wide Web sites which you might find helpful for its resources.

*U*NIQUE FEATURES

A distinguishing feature of this book is its bundling of social trend data from the 1950 through 1990 US censuses which have been carefully compiled to examine significant societal trends, and are centered around the Investigation Topics. Attention is also given to important current issues and social differences that can be investigated best with rich 1990 census data. For example, in Topic Three, students are able to compare the relationship between duration in the US and immigrant assimilation with respect to occupation, earnings, and English language proficiency for specific Latin American- and Asian-origin groups. For Topic Six, they can assess gender inequality in earnings for men and women in specific occupations such as full-time year-round physicians. Because census concepts are nearly universally recognized and stay fairly constant over time, we introduce a *Key Concepts* section for each Investigation Topic which defines in simple terms how concepts like poverty, the family, and labor force are operationalized by social scientists.

StudentChip, the computer software included with this book, was carefully chosen because of my interest in making these investigations readily accessible. Most students will be able to learn the features of StudentChip in one class session. As a further aid, many of the exercises are centered around graphics. Students are asked to translate their tables into line graphs, bar charts or pie charts to facilitate interpretation. An hour spent with our tutorials, *Using StudentChip to Create Tables* and the *Graphing Overview*, along with a little practice, is all that is needed to get going.

*I*NSTRUCTOR SUPPORT AND WORLD WIDE WEB "HOMEPAGE"

We have established a network of faculty members who wish to trade experiences or share exercises using this approach. Funded in part by the Department of Education FIPSE and the NSF Undergraduate Faculty Enhancement program, it is called the Social Science Data Analysis Network (SSDAN), and you can access the World Wide Web "Homepage" which has been established to facilitate conferencing and networking on these topics, and to permit *downloading* of updated datasets at the following address: **http://www.psc.lsa.umich.edu/SSDAN/**. An instructor's manual is also available which provides answers to the exercises and further tips on making this approach both relevant and instructive for your students.

"*J*UST DO IT!"

This brings me back to my "Just do it" philosophy. Based on a decade of firsthand experience here at Michigan and that related to me by dozens of other faculty, I can tell you that it *works* to introduce data analysis exercises into introductory courses, and substantive courses taken by freshmen, sophomores and juniors. By marrying data analysis to engaging substantive questions and issues, students at all levels come to appreciate *why* empirical evidence is important and can actually have fun doing it. I invite your comments, criticisms, and shared experiences about your use of these materials in your class.

ACKNOWLEDGEMENTS

By one way of reckoning, this book has been in the works for almost a decade. In 1987, the University's Provost office awarded me an Undergraduate Initiatives grant to develop a, then, innovative course to introduce data analysis to sociology undergraduates at an early stage. Sociology 231, "Investigating Social and Demographic Change in America," is now a regular offering thanks to the support of the Department of Sociology, the Population Studies Center, and the encouragement of my College's Associate Dean for Undergraduate Education. Along the way, we have been able to disseminate computer materials consistent with this approach to other campuses, first with a grant from the Alfred P. Sloan Foundation. More recently, with a grant from the US Department of Education FIPSE, a formal tie-in was established with the Great Lakes Colleges Association (GLCA), and a more broad-based national distribution was facilitated by a grant from the National Science Foundation Undergraduate Faculty Enhancement program. Together, these dissemination activities have helped to create the Social Science Data Analysis Network discussed in the *Preface*.

No one has been more encouraging toward my efforts than Jim Davis who for years has advocated the practice of integrating data analysis into a variety of undergraduate courses and, through his workshops at Harvard University, forged net-

Back Row: Gary Pupurs, Doug Geverdt, Bridget Fahrland, Ron Lue-Sang, Alexis Miesen, Roland Philbin.
Front Row: Leo Addimando, Lisa Platt, Cheryl First, William Frey, Cathy Sun, Sheila Young.
Not Pictured: Eric Benson, Kristine Pettersen, Katie Richter, Karen Ross.

works among like-minded faculty. Through Jim's network, I have established valuable colleagueships with people like Gregg Carter at Bryant College, Sharlene Hesse-Biber at Boston College, Jere Bruner at Oberlin College, Karen Frederick at St. Anselm College, and Sandy Peterson-Hardt at Russell Sage College, among others. Tony Catanese, of DePauw University, introduced me to another network of colleagues, at the GLCA colleges, who also advanced our efforts.

This book could not have been written without the strong support of Eve Howard, Susan Shook and Deirdre McGill, editors at Wadsworth. Ruth Bogart of Zeta Data has been wonderfully cooperative in helping us bundle StudentChip with our datasets. The preparation of census information for this book's Investigations benefited from the wisdom and assistance of many people at the US Census Bureau including Robert Kominski, Barbara Aldrich, Valerie Gregg, and Dorothy Jackson. I am also grateful for the enormous help I received from the Population Reference Bureau in Washington, DC and thank Peter Donaldson, Kimberly Crews, Carol De Vita and Kelvin Pollard for their efforts.

In the end, the research, background work and production of this book was very much a local team effort (team pictured above). Serving with me as co-leader of this team was Cheryl First MSW, whose managerial wizardry, energy, tenacity and just plain hard work were essential toward turning the initial idea of this book into a reality. Several parts of this book benefited especially from Cheryl's expertise in gender studies and considerable work with women's issues. The datasets, bundled with this book, could not have been produced without the efforts of Cathy Sun, of the Population Studies Center computing staff, who is an expert in census concepts and historical census computer files. Douglas Geverdt, a doctoral student in education, was of great assistance in constructing datasets, conducting background research, and drafting text exercises for several book chapters. Bridget Fahrland did a superb job as our contributing editor. Eric Benson helped to lighten up the pages with his original cartoons. The elaborate page layouts and design were performed by our graphic artists Lisa Platt, Roland Philbin, and Gary Pupurs. I am also grateful to former students in my undergraduate course, Karen Ross and Alexis Miesen as well as Leo Addimando and Kristine Pettersen for their research and contributions to various chapters of this book. Lastly, I am indebted to all the students who have taken Sociology 231 since its first offering in 1987, as their ideas, feedback and comments are reflected in the Investigations and chapters that follow.

William H. Frey
Population Studies Center
and Department of Sociology
University of Michigan

SECTION I
Overview and Getting Started

Investigating Change in American Society

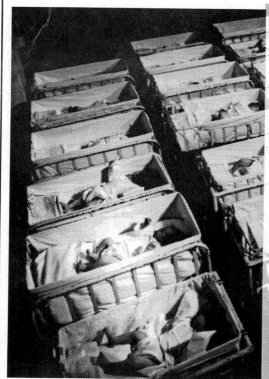

Reuters/Corbis-Bettmann

Reuters/Corbis-Bettmann

1945, May 7 - Germany surrenders, War in Europe is over.

1945, Sep. 2, - Japan Surrenders, end of World War II.

1946 - The Baby Boom begins.

1946 - Dr. Benjamin Spock's *The Common Sense Book of Baby and Child Care* is published.

1947 - Jackie Robinson signs with the Brooklyn Dodgers.

1947 - *Howdy Doody* premieres on TV.

1948 - The McDonald brothers open their first restaurant in San Bernadino, California.

1950 - Korean Conflict begins.

1951 - *I Love Lucy* premieres on TV.

1952 - *The Adventures of Ozzie and Harriet* debuts.

1952, Nov. - Dwight D. Eisenhower elected President.

1952 - Passage of the McCarran Walter Act. Delineates the Asia-Pacific Triangle and establishes a quota system for immigrants from countries in this triangle.

1953, Aug. 7 - Refugee Relief Act (admitted 214,000 more refugees).

1953, - Department of Health, Education and Welfare created by Congress.

1953 - Korean War ends.

1954, May 17 - Brown vs. Board of Education of Topeka, Kansas (segregation in public schools unconstitutional).

1954 - C.A. Swanson and Son introduce frozen dinners.

1954 - Elvis Presley's first professional record is released.

1955 - Bus boycott starts in Montgomery, Alabama.

1955 - Disneyland opens in Los Angeles.

1955 - The song *Rock Around the Clock* begins the rock era.

1955 - Captain Kangaroo is introduced.

1955 - Mickey Mouse Club Show debuts.

1957 - Dr. Seuss publishes *The Cat in the Hat*.

1957 - *Leave It To Beaver* debuts on TV.

1957 - Vietnam War begins.

1957 - President Eisenhower orders federal troops to Little Rock, Arkansas to prevent interference with school integration at Central High School.

1960's - Motown Music becomes popular.

1960, Nov. - John F. Kennedy elected President.

1960 - The first oral contraceptive pill is sold in the U.S.

1962 - Cesar Chavez begins organizing farm workers and goes on to found and preside over the United Farm Workers.

1962 - James Meredith is escorted to campus of the University of Mississippi by Federal Marshals.

1963 - 250,000 march in DC for Civil Rights; Martin L. King, Jr. gives "I have a Dream Speech".

1963, Nov. 22 - John F. Kennedy assassinated, Lyndon B. Johnson sworn in as President.

1963 - *The Feminine Mystique* is published.

1964 - The Baby Boom ends.

1964 - Civil Rights Act of 1964 (prohibited discrimination in public places based on race, color, religion, national origin and gender; also created Equal Employment and Opportunity Commission, EEOC).

1964 - Beatles come to the US.

1965 - President Johnson sends US Marines to Da Nang, Vietnam.

Reuters/Corbis-Bettmann

1965 - Voting Rights Act of 1965 (literacy tests for voter registration made illegal).

1965 - Immigration and Naturalization Act of 1965 (eliminated restrictive race/ethnic quotas).

1965 - Medicaid, Medicare, and Head Start programs begin as part of Great Society's War on Poverty.

1965, Aug 6-11 - Riots in Watts, L.A., and other cities leaving fifty-five dead and 200 million dollars in damages.

1966 - *Star Trek* debuts on TV.

1966 - Janis Joplin's music becomes known as psychedelic music and becomes part of the Hippie counter culture.

1967 - 17,000 Americans died in Vietnam since 1961.

1967 - President Lyndon Johnson appoints Thurgood Marshall to the Supreme Court.

1967, Apr. 4 - Martin L. King, Jr. assassinated.

1968, Nov. - Richard M. Nixon elected President.

1969 - Neil Armstrong is first to land on moon.

1969 - Stonewall Riot - Gays protest police raids and Gay Rights movement is born.

1969 - Woodstock

1969 - *The Brady Bunch* and *Sesame Street* debut.

Reuters/Corbis-Bettmann

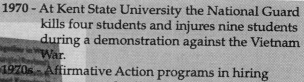

Reuters/Corbis-Bettmann

1970 - At Kent State University the National Guard kills four students and injures nine students during a demonstration against the Vietnam War.

1970s - Affirmative Action programs in hiring contracts and college admissions actively pursued.

1971 - *All in the Family* and The *Sonny and Cher Comedy Hour* debut on TV.

1972 - 27th Amendment a.k.a. the Equal Rights Amendment passes Senate and is given to states for ratification.

1972 - *Ms.* magazine begins publication.

1972, Jun. 17 - Watergate Affair disclosed.

1973 - OPEC oil embargo begins.

1973-75 - Economic recession.

1973 - The last US ground troops leave Vietnam.

1973 - Roe vs. Wade (US. Supreme Court legalizes women's right to abortion).

1974, Aug. 9 - Richard M. Nixon resigns from office .

1974 - *Happy Days* debuts on TV.

1975 - South Vietnam surrenders to North Vietnam.

1975 - Indian Self-Determination and Education Assistance Act strengthens the ability of tribal governments to govern federally funded programs themselves.

1975 - NBC's *Saturday Night Live* debuts.

1975 - *Jaws* is released and is the top moneymaker movie of the year.

1975 - Disco arrives.

1975 - First personal computer for home use is introduced by retailers.

1976, Nov. - James E. Carter elected President.

1976 - The *Muppet Show* and the *Donnie and Marie Show* debut.

1976 - Apple computer is developed in a California garage.

1977 - Elvis Presley dies.

1977 - *Star Wars* and *Saturday Night Fever* are released.

1978 - *Bakke vs. University of California* (affirmative action, yes, decisions based on quotas, no).

1979 - Sony Walkman Radio debuts in the United States.

Reuters/Corbis-Bettmann

1980, Nov. - Ronald W. Reagan elected President.

1981 - Sandra Day O'Conner becomes the first women to be confirmed to the United States Supreme Court.

1981 - Dr. Ruth radio talk show debuts.

1981 - Music Television (MTV) debuts.

1982 - Equal Rights Amendment still not ratified by the necessary 38 states.

1982 - HIV/AIDS emerges as an increasing cause of death in the U.S.

1982 - *E.T.* is released.

1982 - Michael Jackson releases *Thriller* which becomes the top selling album in history.

1983 - Asian Immigrant Women Advocates (AIWA) was founded.

1984 - Geraldine Ferraro becomes first women nominated for the Democratic Vice-Presidential ticket.

1984 - Berkeley becomes the first U.S. city to extend spousal benefits to live-in partners of gay and lesbian employees.

1985 - Crack cocaine appears in the U.S.

1986 - Drexel, Barnham, Lambert executive Dennis Levine pleads guilty to insider trading.

1986, Nov. 6 - Immigration Reform and Control Act (IRCA), legalized some illegal aliens, strengthened Border Patrol and forbade U.S employers from hiring illegal aliens.

1987 - On October 19, the Dow drops 508 points. The day is remembered as Black Monday.

1988, Nov. - George H. Bush elected President.

1988 - End of publicly funded abortions.

1989 - Liu Gang leads student protests in Tiananmen Square, China.

1989-93 - Economic Recession.

1990, Aug. 2 - Iraqi troops invade Kuwait.

1990 - *Seinfeld* debuts on TV.

1991, Apr. 6 - Iraqis accept cease-fire agreement and Persian Gulf War ends.

1992, Nov. - William H. Clinton elected President.

1992, Apr. - Riots in L.A. and elsewhere because of Rodney King trial decision.

1993 - Family Leave Act of 1993.

1994 - Proposition 187 initiative approved in California (denies illegal aliens public health, education and social services).

1995 - Violence Against Women Act passed.

1996 - First Baby Boomer turns 50.

Reuters/Corbis-Bettmann

Reuters/Corbis-Bettmann

*I*NVESTIGATING CHANGE

America has undergone some fairly dramatic changes in just the last several decades. The 1950s were a stable decade with a booming economy and with *Leave It to Beaver* families as the norm. Still, a new brand of music with the likes of Elvis Presley and rock-n-roll hinted that more socially rebellious times were on the way. The term "rebellious" is certainly an appropriate one to characterize the 1960s. During the 60s, a new social awareness regarding issues of civil rights for minorities and women led to new legislation and increased federal spending on social programs. Perhaps the most rebellious people of this period were the early *baby boomers*, many in college, who protested against the conventions of established society and became especially energized in their feelings about the Vietnam war.

Although these social concerns, including the women's movement and Gloria Steinem's *Ms. Magazine*, continued through the next decade, the '70s were a period when less idealistic political and economic realities set in. The Watergate scandal led to the resignation of President Richard Nixon and this was soon followed by a severe economic recession just as large numbers of baby boomers, minorities and women began entering the labor force. The "traditional family" also began to disintegrate with later marriages, lower fertility, and increasing rates of divorce.

In the 1980s, America's political inclinations became much more conservative with the election of Ronald Reagan and the ascendancy of more traditional religious groups. With these changes, some social and economic divisions became even wider. Debates over issues like abortion rights and nuclear proliferation became more intense. Huge stock market gains were made, but poverty was also on the rise and the middle class began to shrink.

In some ways the 1990s brought greater moderation. Young adult *Generation Xers* are more technologically literate than earlier cohorts and also more practical about what they can accomplish in light of new opportunities and changing economic conditions. The first baby boom presidential ticket of Bill Clinton and Al Gore presided over a government that was neither as ambitious as Lyndon Johnson's *Great Society* in the 1960s, nor as conservative as the administration of Ronald Reagan in the 1980s. Yet, strong social divisions still exist along with new issues associated with the large immigrant Latino and Asian populations as they assimilate into American society. Affirmative action as a solution toward resolving inequities among different racial and gender groups has come into question, and the divergent economic circumstances of America's children versus its older population has become a new concern.

𝒫URPOSE OF THIS BOOK

The remarkable changes which have occurred over the last several decades have left their *imprint* on today's society and population. Social differences which exist today between men and women, and across race/ethnic groups, generations, and social classes can best be understood by tracking their progression over the period just described. By examining these trends, you can interpret how these differences have evolved in response to important social, economic and political events.

Take, for example, the impact of the Civil Rights legislation of the 1960s. That legislation helped to narrow the gaps between blacks and whites in terms of educational opportunities, job access and earnings potential. Yet this legislation did not have an immediate impact on all of society. Rather, its effect was gradual, both over time and for particular generations. So, when we examine these changes over time, you will find that the blacks in the baby boom generation were really the first to take full advantage of the Civil Rights movement in terms of social advancement, and this has continued with the later *Generation Xers*. The point is that when you look at the total population today, social class differences between blacks and whites—while they still exist—are much narrower among those below the age of 50 because the Civil Rights *imprint* was stronger on these generations.

The materials in this book will allow you to conduct *hands-on* investigations of data to explore the wide range of issues associated with recent changes in America's population. In the process, you will come to know a lot more about the differences that exist within our society today.

𝓗ANDS-ON EXPLORATIONS WITH U. S. CENSUS DATA

The most comprehensive data set needed to assess the kinds of over-time changes that we have been discussing are available from the U. S. Census (see *What is the Census?* at the end of this book). Normally, social scientists have to spend hours in the library working with published census volumes, or CD-ROM's, in an attempt to mine these valuable statistics. The good news for you is that we have done all this work ahead of time. A diskette, bundled with this book, contains an enormous amount of census data specially tailored to answer all of the questions, and many more, contained in the investigations that follow. This will allow you to immediately get your *hands on* the data, to explore the question, and then to investigate it even further with the wide range of additional relevant data items we have included.

The other piece of good news for you is that the computer program bundled with this book, *StudentChip* (a student version of the popular Chipendale software), is the most user-friendly software we could find. Most students, even those who have never used a computer or done anything mildly statistical, can learn all there is to know about this program in an hour. Of course, accessing the data with this program is not the same as learning how to analyze it and your instructor will help there. However, just *getting into* the program will be fairly easy. We have provided a simple tutorial, but many students find that they do not need this because the program is so intuitive. Either way, you will find that exploring these U. S. census data will not be a chore, and can actually be fun.

*H*APPY INVESTIGATING!

This book was written to allow you to explore a whole range of social and economic issues that are facing society today. Explorations associated with racial inequality, immigrant assimilation, gender inequality and the *glass ceiling*, as well as differences in poverty levels among children and the elderly represent a few of the topics covered. The fact that your instructor has assigned this book means that he or she is interested in having you get a taste for how social scientists begin to explore evidence in answering questions that are current and relevant.

After you get this taste, you will undoubtedly come back to the datasets and do further explorations. We have included some aids to help you along. Each new topic provides a section called *Key Concepts*. These are intended to give you a simple definition of the social or economic concepts (such as poverty or cohabitation) introduced. At the end of the book are a number of useful resources, including a guide to all datasets, an index to the *Key Concepts*, related readings, and some resources on the World Wide Web. So in addition to your class assignments, this book also provides you with do-it-yourself tools for exploring topics with the data sets.

It may be that even after you've finished the course, you will become *hooked* on the habit of exploring questions about social trends and differences with the data sets that we provide or others you may seek out. Even if you don't, this book will arouse your curiosity, teach you something about American society, and give you a feel for the way social scientists do their work. Happy Investigating!

$\mathcal{U}$SING *StudentChip* TO CREATE TABLES–

*I*NTRODUCTION

This tutorial illustrates how to use the StudentChip program to analyze our SSDAN census datasets with table analysis (crosstabs) and a few other features. It is not meant to be an exhaustive explanation of StudentChip, but will walk you through the steps required to answer basic questions using the program. This tutorial is broken into the following sections: Accessing StudentChip and Finding a Dataset, Saving Your Output, Cross Tabs, Modifying a Variable, and Exiting the Program.

The example we will use here focuses on the following questions: Are there differences in the earnings of women and men, ages 35-44, in 1990? Do these differences decrease when we look at specific occupations? To investigate these questions we will use the dataset **WORK9-35.DAT**, taken from the 1990 U.S. Census, containing information on women and men ages 35-44 who work full time, year round.

*A*CCESSING STUDENTCHIP AND FINDING A DATASET

To investigate the above question, we need to, first, access StudentChip, and second, locate the dataset **WORK9-35.DAT**.

To access StudentChip: (1) Put the disk in the drive (2) Type a: and press enter on the keyboard, and (3) type chip after the a:\> prompt. The screen will begin to blink as the program is opening.

Now locating dataset **WORK9-35.DAT** requires knowing how datasets are organized. The datasets on your disk are all contained in the directory **FREYCEN(DIR)** which is subdivided into two subdirectories. The first is titled **CENTrend**, and it contains trend datasets usually spanning the years 1950 to 1990. The second, titled **CEN1990**, contains datasets using 1990 census data. A complete list of all datasets in both subdirectories can be found in the *Guide to Datasets* at the end of this book. However, you can always tell a dataset in the **CEN1990** subdirectory because it has a single "9" in the name (Datasets in the CENTrend subdirectory usually have a "5090" or "7090" in the name). Therefore, **WORK9-35.DAT** is in the **CEN1990** subdirectory, so we will proceed to locate it there.

Once you are in the StudentChip program, you will see a screen like this:

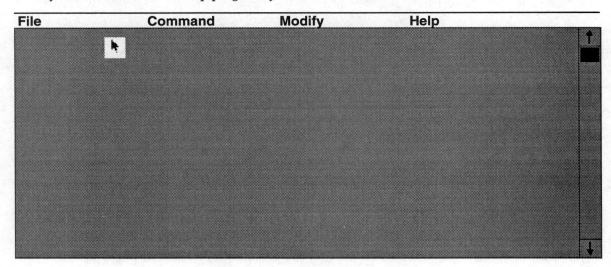

To find a dataset, click on the FILE menu, and select the OPEN command. Your screen will now look like this:

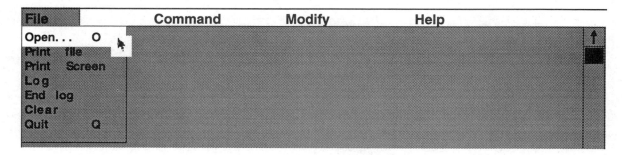

Use your mouse to highlight **FREYCEN (DIR)** and then select the "OK" button . Your screen will then look like this:

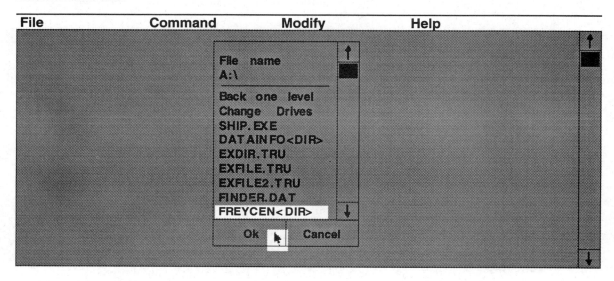

Use your mouse to first highlight **CEN1990 (DIR)** and then to select the "OK" button. Your screen will then look like this:

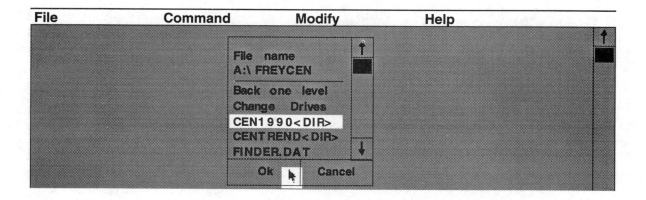

Next, point your mouse to the down arrow key, on the lower right side of the file name screen, and press it until you have scrolled down far enough where **WORK9-35.DAT** is visible. Then highlight this file and select "OK" with your mouse. Your screen will now look like this:

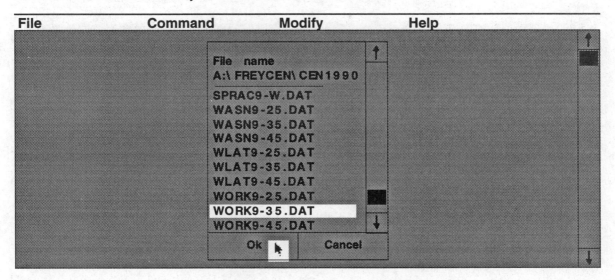

StudentChip will display basic information about the dataset **WORK9-35.DAT**, including its selected population: 1990 Full-time, Year Round civilian workers, and the size of the population. Your screen will now look like this:

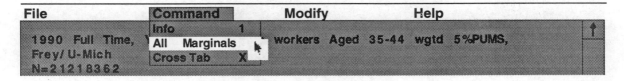

You can find out about the variables in the dataset **WORK9-35.DAT** by clicking your mouse on the command menu and selecting ALL MARGINALS under the COMMAND menu (shown below).

File	Command	Modify	Help
	Info 1		
1990 Full Time,	All Marginals	workers Aged 35-44 wgtd 5%PUMS,	
Frey/ U-Mich	Cross Tab X		
N=21218362			

The ALL MARGINALS option displays all of the variables and categories included in the dataset. After you have selected the ALL MARGINALS option, here is what the output should look like for the dataset **WORK9-35.DAT** (as shown on next page).

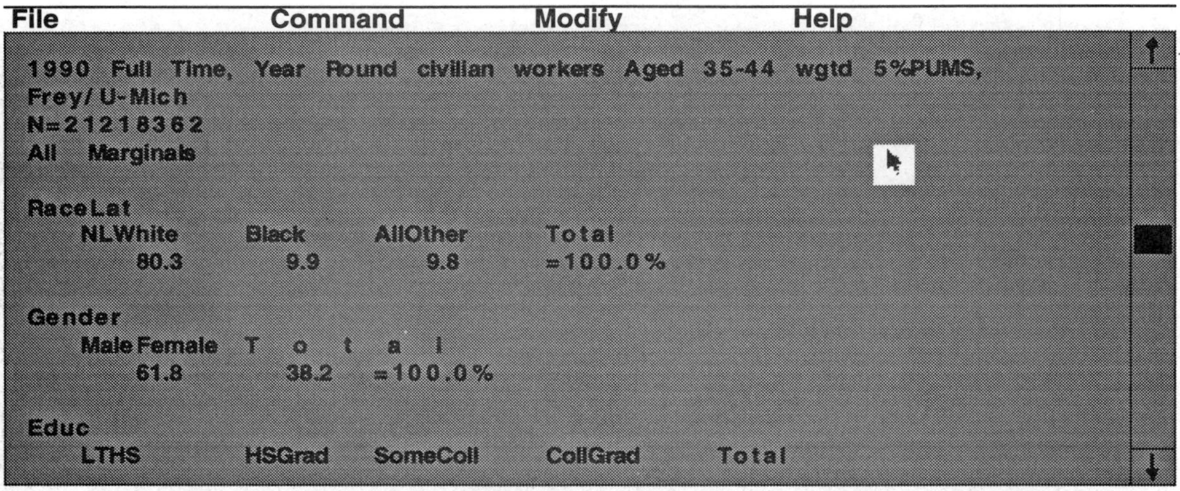

File	Command	Modify	Help	

1990 Full Time, Year Round civilian workers Aged 35-44 wgtd 5%PUMS,
Frey/ U-Mich
N=21218362
All Marginals

RaceLat

NLWhite	Black	AllOther	Total	
80.3	9.9	9.8	=100.0%	

Gender

Male	Female	T o t a l	
61.8	38.2	=100.0%	

Educ

| LTHS | HSGrad | SomeColl | CollGrad | Total |

The marginals show that dataset **WORK9-35.DAT** contains five variables: RaceLat (race Latino status), Gender, Educ (education), Occup (occupation) and Earnings. In order to see Occup and Earning, use your mouse to click on the down arrow key (as shown below).

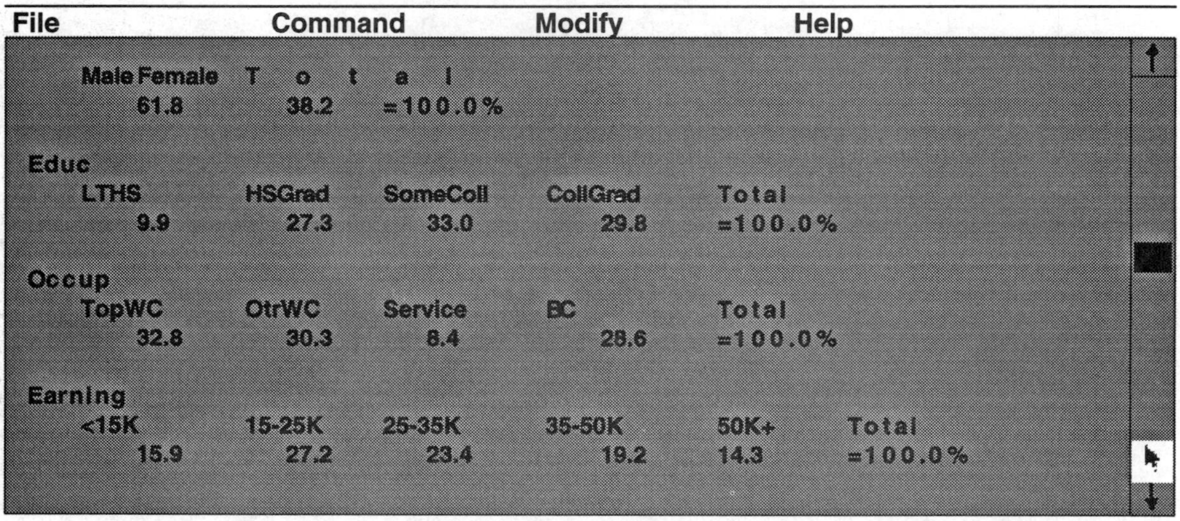

File	Command	Modify	Help	

Male	Female	T o t a l		
61.8	38.2	=100.0%		

Educ

LTHS	HSGrad	SomeColl	CollGrad	Total
9.9	27.3	33.0	29.8	=100.0%

Occup

TopWC	OtrWC	Service	BC	Total
32.8	30.3	8.4	28.6	=100.0%

Earning

<15K	15-25K	25-35K	35-50K	50K+	Total
15.9	27.2	23.4	19.2	14.3	=100.0%

Since we are mainly interested in gender, earnings, and occupation, we will focus on these particular variables for our tables. Note that the earning variable contains five categories: <15K (less than $15,000), 15-25K ($15-25,000), 25-35K ($25-35,000), 35-50K ($35-50,000), and 50K+ (above $50,000). Note that 14.3% of the people in this analysis earn more than $50,000 per year.

The occupation variable contains four categories (see picture above): TopWC (Top White Collar), OthWC (Other White Collar), Service, and BC (Blue Collar). Note that 32.8% of the people are employed in top white collar jobs, which includes managers and professionals.

SAVING YOUR OUTPUT

Before you proceed further with analysis, you need to make an output file. The reason that this is important, is so that when you are done using StudentChip, you will want to have a file containing all of your work. To make an output file, go back and pull down the FILE menu and select the LOG option (as pictured on the following page).

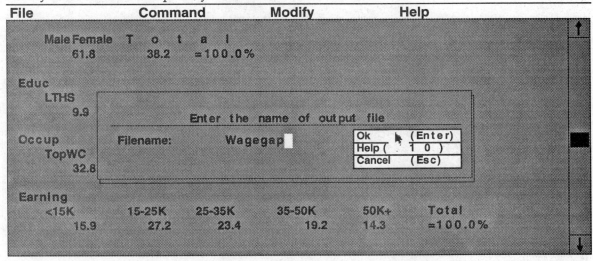

File		Command	Modify		Help		
Open... O		T o t a l					
Print file		38.2	=100.0%				
Print Screen							
Log ▸							
End log		HSGrad	SomeColl	CollGrad	Total		
Clear		27.3	33.0	29.8	=100.0%		
Quit Q							
Occup							
TopWC	OtrWC	Service	BC	Total			
32.8	30.3	8.4	28.6	=100.0%			
Earning							
<15K	15-25K	25-35K	35-50K	50K+	Total		
15.9	27.2	23.4	19.2	14.3	=100.0%		

After you select the LOG option, your screen will look like this:

File		Command	Modify	Help	
Male Female T o t a l					
61.8	38.2	=100.0%			
Educ					
LTHS					
9.9					
		Enter the name of output file			
Occup	Filename:	Wagegap		Ok ▸ (Enter)	
TopWC				Help (1 0)	
32.8				Cancel (Esc)	
Earning					
<15K	15-25K	25-35K	35-50K	50K+	Total
15.9	27.2	23.4	19.2	14.3	=100.0%

You will be asked to enter a short file name of your choice of either letters or numbers. We have named our output file "Wagegap." Please name your output file and then select the "OK (Enter)" button (also remember the name you gave to your output file!).

After you are done with your work, but before you exit the program go back to the File menu and select END LOG. This will save the output file, with your chosen name, on your disk. You can later access this file with any word processing program (like Microsoft Word or WordPerfect) or spreadsheet (like Excel) to observe the results of your analysis. You can also print your output file immediately by using your mouse to access the PRINT FILE command which is also found under the FILE menu.

MAKING A CROSSTAB

Returning to our example, we want to use the dataset **WORK9-35.DAT** to compare men's earnings with women's earnings. This requires making a table which cross tabulates GENDER with EARNING. To do this, we will construct a table using the CROSSTAB command located under the COMMAND menu (see picture next page).

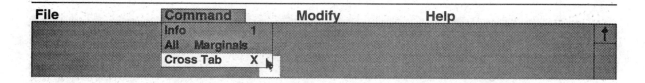

Use your mouse to highlight the variable GENDER, select the OK button, then highlight the variable EARNING, and select the OK button once more. Your screen should now look like this:

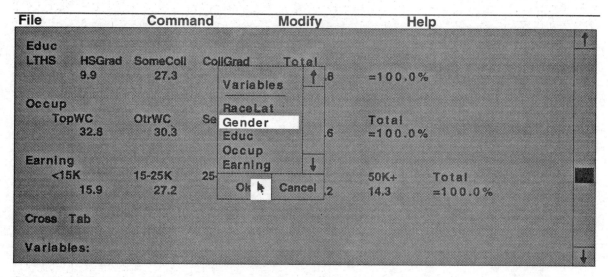

(Notice: the first variable you select, GENDER , will be the rows of the table, and the second variable, EARNING, will be the columns of the table.)

Then pull down the OPTIONS menu and select PERCENT ACROSS (see picture below).

Options		Standard			Help		
Frequency	neColl	CollGrad	Total				
Percent across	27.3	33.0	29.8	=100.0%			
Percent down							
Percent diff							
Chi square	WC	Service	BC	Total			
Control C	30.3	8.4	28.6	=100.0%			
Release control R							
Print screen							
Plot	25K	25-35K	35-50K	50K+	Total		
Clear	27.2	23.4	19.2	14.3	=100.0%		
Exit E							
Cross Tab							
Variables:							
Gender/ Earning							

(Note: in the exercises, you will sometimes need to use the PERCENT DOWN function which is also located in the OPTIONS menu, directly below the PERCENT ACROSS option).

After selecting the PERCENT ACROSS option, your screen will look like this:

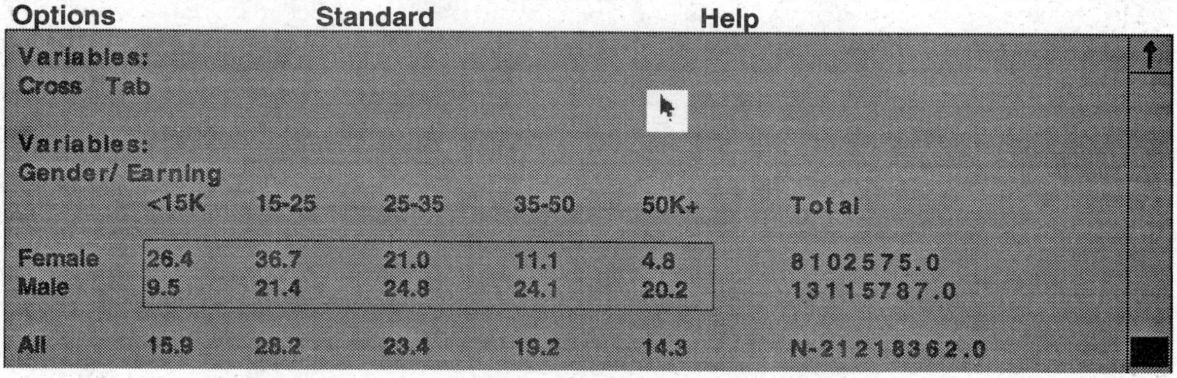

Options		Standard			Help	

Variables:
Cross Tab

Variables:
Gender/ Earning

	<15K	15-25	25-35	35-50	50K+	Total
Female	26.4	36.7	21.0	11.1	4.8	8102575.0
Male	9.5	21.4	24.8	24.1	20.2	13115787.0
All	15.9	28.2	23.4	19.2	14.3	N-21218362.0

The above table shows that of all women, 26.4% earn less than $15,000, 36.7% earn between $15,000 and $25,000, and so on. Note that all of these earnings categories sum to 100%. The "total" category shows the number of women in the sample. Of men, 9.5% earn less than $15,000, and so on. The "All" row represents the percentage of the entire sample in that earnings category. By analyzing this table, we see that there are actually large differences between men and women in earnings. A larger percentage of women earn less than $15,000 while a much smaller percentage earn more than $50,000; only 4.8% of women, compared to 20.2% of men, earn more than $50,000 per year.

Once you are in StudentChip, there are two levels of menus, which you will notice as you progress through the program. The first level, which we used earlier, allows you to open files, get basic file information like marginals, modify variables, and other basic tasks. It includes: FILE, COMMAND, MODIFY, and HELP. The second level appears while you are doing crosstabs and includes: OPTIONS, STANDARD, and HELP.

While in the second level, you can return to the first level by selecting the EXIT command from the OPTIONS menu, which ends your work within this crosstab and takes you back to the first level of menus. From the first level, you can select a new crosstab with this dataset, open a new dataset (from the FILE menu as discussed earlier) or exit the program by selecting EXIT from the FILE menu.

Using the "Frequency" Option

Another function that you will be asked to use, is the FREQUENCY function. This function allows you to replace the percentages in our PERCENT ACROSS table (pictured above), with the actual number of people that this data file had under that category. If you were to go back to the OPTIONS menu, pictured on the previous page, and select the FREQUENCY option, your screen would look like this:

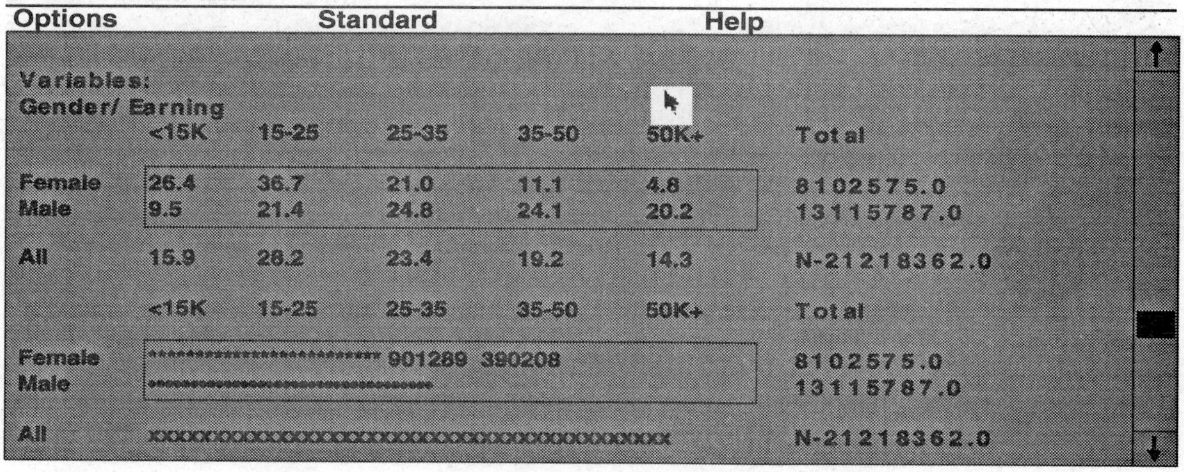

Options		Standard			Help	

Variables:
Gender/ Earning

	<15K	15-25	25-35	35-50	50K+	Total
Female	26.4	36.7	21.0	11.1	4.8	8102575.0
Male	9.5	21.4	24.8	24.1	20.2	13115787.0
All	15.9	28.2	23.4	19.2	14.3	N-21218362.0
	<15K	15-25	25-35	35-50	50K+	Total
Female	******************* 901289 390208					8102575.0
Male	••••••••••••••••••••••••••					13115787.0
All	xxxxxxxxxxxxxxxxxxxxxxxxxxxxxxxxxxxxxx					N-21218362.0

You will notice that there are many asterisks where you would expect to find numbers. This is due to the lack of space in that small area to fit all the necessary numbers. In the exercises contained in this book, the numbers will usually be small enough so that there is sufficient room for them to be displayed on the screen.

CROSSTABS THAT "CONTROL" FOR ADDITIONAL VARIABLES

Instead of leaving the crosstab, suppose we want to investigate gender differences in earnings further. Therefore, remain with the second set of menus. We want to see if our observed differences in earnings, between men and women, depend on particular occupations. Therefore we want to compare the earnings of women and men within specific occupations. We call this "controlling" the original table for occupation. In effect, we are looking at earnings differences between men and women under the "controlled" conditions that they have the same occupations. To do this, you simply go to the OPTIONS menu and select CONTROL (see picture below).

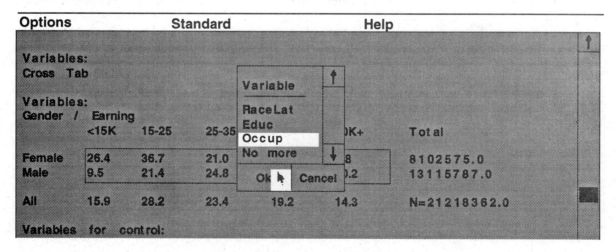

Options		Standard		Help	
Frequency					
Percent across					
Percent down					
Percent diff	-25	25-35	35-50	50K+	Total
Chi square					
Control C	.7	21.0	11.1	4.8	8102575.0
Release contro. R	.4	24.8	24.1	20.2	13115787.0
Print screen					
Plot	.2	23.4	19.2	14.3	N-21218362.0
Clear					
Exit E	-25	25-35	35-50	50K+	Total

From there, you use your mouse to highlight the variable OCCUP, and then select "OK." Finally, use your mouse to highlight NO MORE, and then select the "OK" button once more. Your screen will now look like this:

Options			Standard			Help		
Variables:								
Cross Tab				Variable				
Variables:				RaceLat				
Gender /	Earning			Educ				
	<15K	15-25	25-35	Occup		)K+	Total	
				No more				
Female	26.4	36.7	21.0			8	8102575.0	
Male	9.5	21.4	24.8	Ok Cancel		.2	13115787.0	
All	15.9	28.2	23.4	19.2	14.3		N=21218362.0	
Variables for control:								

You then return to the OPTIONS menu and select PERCENT ACROSS again, and this is what you will see (see next page):

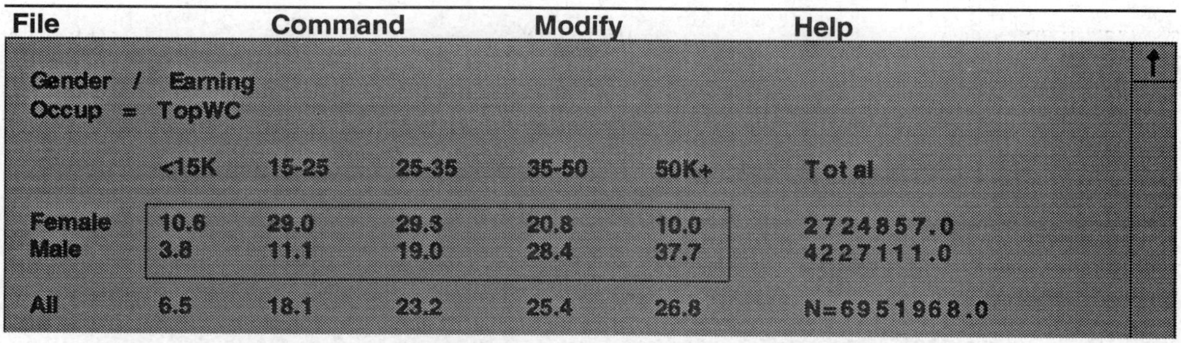

Gender / Earning Occup = TopWC	<15K	15-25	25-35	35-50	50K+	Total
Female	10.6	29.0	29.3	20.8	10.0	2724857.0
Male	3.8	11.1	19.0	28.4	37.7	4227111.0
All	6.5	18.1	23.2	25.4	26.8	N=6951968.0

This is one of four tables you will get in this example. The first one shows you the earnings specifically for women and men who are in all top white collar jobs (TopWC). What this table tells us is that within this occupational category, while 37.7% of men earn more than $50,000, only 10% of women earn that much. You would continue by reading the table in the same way as the earlier example.

By clicking the mouse on the down arrow key (located in the bottom right hand corner), or by pressing the RETURN (ENTER) key you can scroll through your new tables that are now specific to people in each occupational category. The tables show the earnings distribution of people with other white collar jobs (OtrWC), then service workers, then blue collar workers (BC).

You could also control for additional variables by selecting more control variables. (For example, if you wanted to control for education in addition to occupation, you would get a table for each combination of education and occupation, totaling 16 tables in all.) Before proceeding, go to the OPTIONS menu and select EXIT. This will take you back to the first level of menus (FILE, COMMAND, MODIFY, HELP).

MODIFYING A VARIABLE

In our example, we used five categories of the EARNING variable, but it is also possible to modify any of the variables included in a data file. By modify we mean that you can combine categories of a variable to make things more simple, or omit categories you don't want.

For example, we might want to combine the two upper categories of the earning variable so that the upper category is "above $35,000". To do this you would need to combine 35-50K and 50K+ into a single category 35K+. In order to do this, go to the MODIFY menu and select the COMBINE option (see picture below).

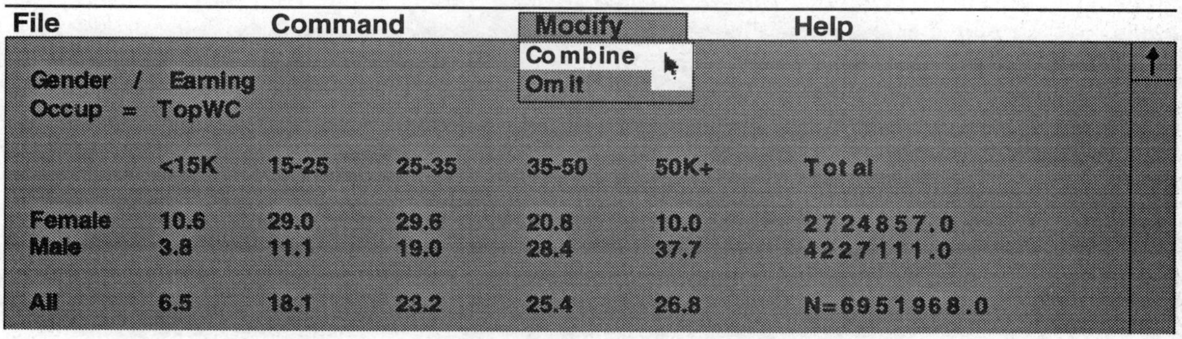

Combine
Omit

Gender / Earning Occup = TopWC	<15K	15-25	25-35	35-50	50K+	Total
Female	10.6	29.0	29.6	20.8	10.0	2724857.0
Male	3.8	11.1	19.0	28.4	37.7	4227111.0
All	6.5	18.1	23.2	25.4	26.8	N=6951968.0

(Note: the process for omitting a category isn't illustrated here, but is much the same as the process of combining categories. Instead of selecting the COMBINE option under the MODIFY menu, you would select the OMIT option. After that, you would follow the same steps listed below for combining.)

Now select the variable you want to modify (for our example, choose EARNING), by highlighting it with your mouse, and then selecting the "OK" button. After you do this, your screen will now look like this:

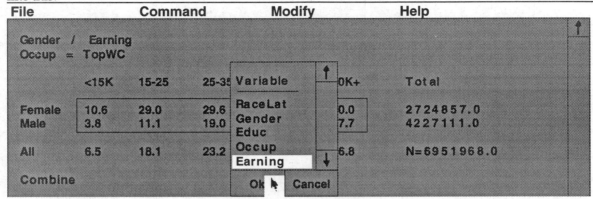

Then select the categories you want to combine by, once again, highlighting them and then clicking the OK button (For our example, select 35-50K and 50K+). When you are done selecting the categories to be combined, highlight the NO MORE option and select OK. Your screen will now look like this:

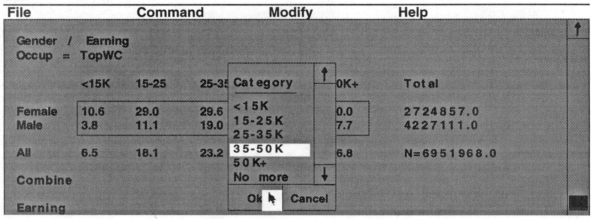

You will be asked to give your new category a label. We have chosen "35+" for our combined category's name. You should choose a name that you will remember. After you have done this select the OK (Enter) button. Your screen will now look like this:

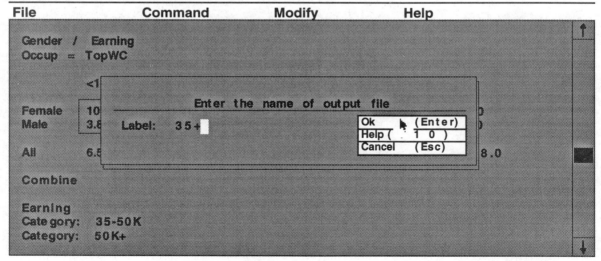

You can now use the ALL MARGINALS option under the COMMAND menu to check the results of your modifications. In order to see the results of your modifications, you will have to look to the last marginal category- EARNING (see picture below). There you should notice that the last two categories have been combined into one called "35+."

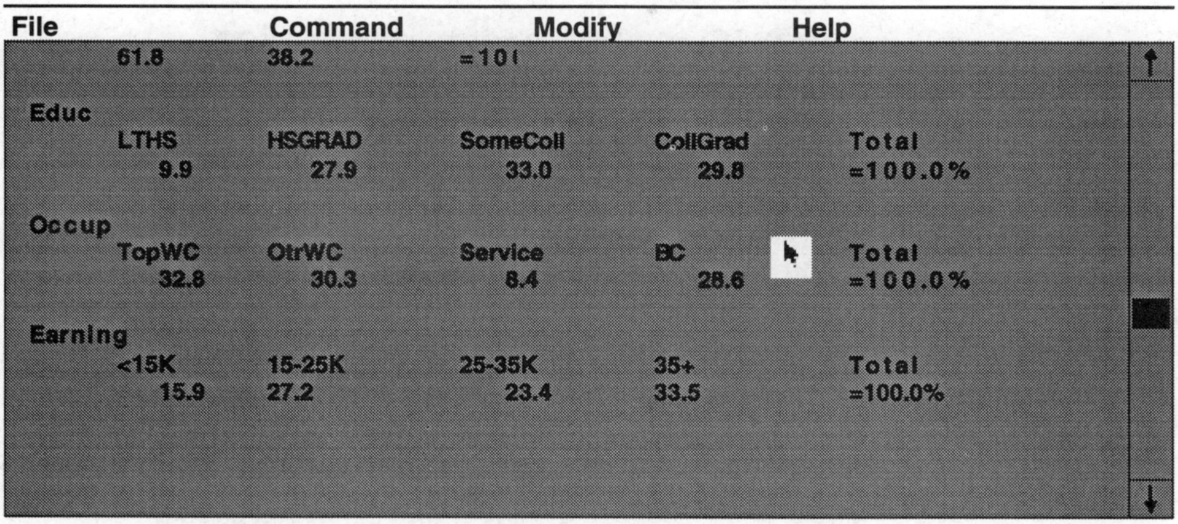

File		Command		Modify		Help	
61.8		38.2		=10(			
Educ							
LTHS		HSGRAD		SomeColl		CollGrad	Total
9.9		27.9		33.0		29.8	=100.0%
Occup							
TopWC		OtrWC		Service		BC	Total
32.8		30.3		8.4		28.6	=100.0%
Earning							
<15K		15-25K		25-35K		35+	Total
15.9		27.2		23.4		33.5	=100.0%

EXITING THE PROGRAM

After you are done using StudentChip, all you need to do in order to exit is return to the FILE menu and select the QUIT option.

GRAPHING OVERVIEW

INTRODUCTION

Graphing skills provide a useful way to highlight important trends and comparisons that are not always evident by looking at numbers and percentages. These skills are not only useful for the exercises contained in this book, but also necessary for reading and understanding many textbooks, periodicals, and other printed materials that contain visual representations of data.

A graph is defined as a pictorial representation (diagram) of the way in which one statistic changes over time, or differs across groups. Graphic representations allow otherwise confusing data to be displayed in a way that is easy to understand.

The purpose of this overview is to familiarize you with the four types of graphs, a line graph, a bar chart, a pie chart, and a stacked bar chart, that you will be asked to work with in the chapters that follow. Additionally, this section will explain how to most effectively display data and illustrate how to transform the tables you produce in the StudentChip program into graphs.

LINE GRAPHS

Line graphs are used to examine the trends for a statistic over time. Line graphs can also allow you to compare the trends of two or more different groups on the same graph so that any similarities/differences can be identified.

Example 1 For this example, use the StudentChip data below to create a line graph showing the percentage of blacks and nonblacks who were never married, from 1950 to 1990.

DATA (*MARR5090.DAT*): Cross Tab= Race / Marital (control for year); Percent Across

Year = 1950

	CurMr	Widow	Divor	Sepra	NevMr	Total
Black	57.6	10.5	2.3	7.8	21.7	100%
NonBl	67.5	8.0	2.2	1.2	21.0	100%
All	66.6	8.2	2.2	1.9	21.1	100%

Year = 1960

	CurMr	Widow	Divor	Sepra	NevMr	Total
Black	56.0	10.0	3.2	7.7	23.2	100%
NonBl	68.7	7.8	2.5	1.2	19.8	100%
All	67.5	8.0	2.6	1.8	20.1	100%

Year = 1970

	CurMr	Widow	Divor	Sepra	NevMr	Total
Black	49.1	9.8	4.3	7.6	29.1	100%
NonBl	64.8	8.0	3.3	1.3	22.6	100%
All	63.2	8.2	3.4	1.9	23.3	100%

Year = 1980

	CurMr	Widow	Divor	Sepra	NevMr	Total
Black	39.4	8.6	7.7	7.3	37.0	100%
NonBl	60.2	7.5	6.0	1.6	24.6	100%
All	58.0	7.6	6.2	2.1	26.0	100%

Year = 1990

	CurMr	Widow	Divor	Sepra	NevMr	Total
Black	35.2	8.0	10.1	6.6	40.1	100%
NonBl	58.1	7.3	8.1	1.7	24.7	100%
All	55.6	7.4	8.3	2.3	26.4	100%

The answer can be plotted as follows:

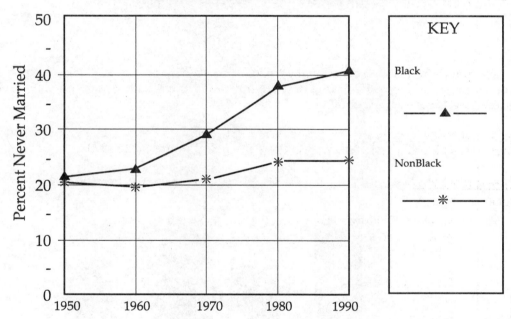

Blacks and NonBlacks Never Married 1950 to 1990

By placing both lines on the same axis, you can:

■ Effectively illustrate the trends, over time, for each group;
■ View the differences in trends between the two groups.

$\mathcal{B}$AR CHARTS

A bar chart is an effective way of showing how groups may differ on a certain statistic. Bar charts are effective because they allow you to more precisely illustrate the exact values of the factor being compared across the observed groups.

Example 2 For this example, use the StudentChip data to illustrate the percentage of people never married by race/ethnicity, in 1990.

DATA (*MARITAL9.DAT*): Cross Tab: RaceLat / Marital; Percent Across

	CurMr	Widow	Divor	Sepra	NevMr	Total
NLOth	42.6	4.2	8.5	3.3	41.3	100%
AmInd	46.6	5.5	11.7	3.5	32.7	100%
Latin	51.5	3.8	7.2	3.8	33.7	100%
Asian	58.6	4.3	3.8	1.5	31.8	100%
Black	35.2	8.0	10.1	6.6	40.1	100%
NLWhi	58.9	7.8	8.3	1.5	23.5	100%
All	55.6	7.4	8.3	2.3	26.4	100%

The answer can be plotted as follows:

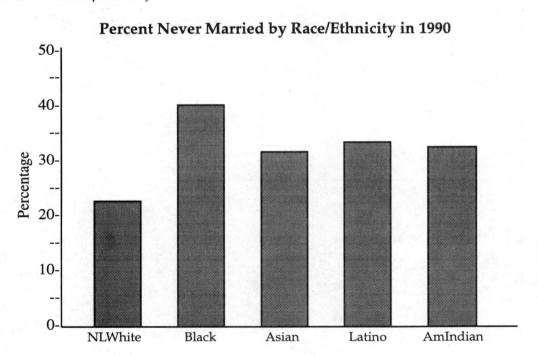

Percent Never Married by Race/Ethnicity in 1990

By taking this data and placing it in the form of a bar chart, you can:

■ Illustrate the specific values for each group;
■ Compare the differences between groups.

$\mathcal{P}$IE CHARTS

The purpose of a pie chart is to show how a population is distributed across all of the categories of a variable, rather than showing just a single statistic from the group. Each *slice* represents a category and the size of the slice is the precentage share of the group. Pie charts allow you to visualize the entire distribution of a variable.

Example 3 For this example, use the data below to chart the marital status of the entire population in 1990.

DATA: (*MARITAL9.DAT*): Marginals

CurMrrd	Widowed	Divorced	Seprated	NevMrrd	Total
55.6	7.4	8.3	2.3	26.4	100.00%

The answer can be graphed as follows:

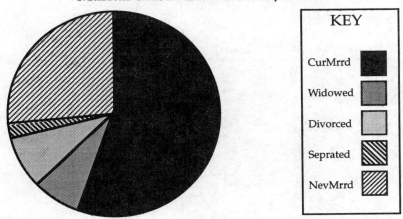

Marital Status Distribution, 1990

By taking this data and presenting it in the form of a pie chart, you can:

■ Visualize the relative percentage of a population that fits into a certain criteria;
■ Compare these relative percentages for different populations.

STACKED BAR CHARTS

A stacked bar chart is used to compare distributions across groups. They provide an alternative to using several bar or pie charts to illustrate and compare groups' distributions of a certain variable. Each bar in a stacked bar chart will represent 100% of a group's population, and all the divisions of the bars represent categories of the variable, acting much like the slices of a pie chart.

Example 4 Use the following data to create a stacked bar chart showing the marital status distributions of Whites, Blacks, Asians, and Latinos.

DATA (_MARITAL9.DAT_): Cross Tab: Race/Marital; Percent Across

	CurMr	Widow	Divor	Sepra	NevMr	Total
NLOth	42.6	4.2	8.5	3.3	41.3	100%
AmInd	46.6	5.5	11.7	3.5	32.7	100%
Latin	51.5	3.8	7.2	3.8	33.7	100%
Asian	58.6	4.3	3.8	1.5	31.8	100%
Black	35.2	8	10.1	6.6	40.1	100%
NLWhi	58.9	7.8	8.3	1.5	23.5	100%
All	55.6	7.4	8.3	2.3	26.4	100%

The answer can be graphed as follows:

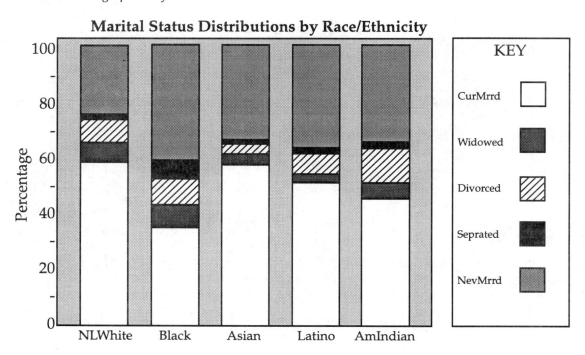

By taking the data and organizing it in the form of a stacked bar chart, you can:

■ Illustrate the distributions within each group;
■ Compare these distributions to those of other groups.

SECTION II
Investigation Topics

Investigating Change in American Society

POPULATION STRUCTURE: COHORTS, AGES, AND CHANGE
topic one

The population structure of the United States will be discussed in each succeeding chapter of this book. Broadly speaking, a population's structure includes its race/ethnicity distribution, its labor force characteristics, families, households, and other social dimensions. The most fundamental aspects of a population's structure are its age and gender structure. These two factors provide the framework for understanding other aspects of population change.

In this chapter, you will become familiar with the age structure of the population and how it changes. Birth cohorts are an important "engine" of change in America's age structure. These cohorts are usually very different sizes. As these cohorts age, the proportion of different age groups in the population changes. The best example is the baby boom cohorts. These huge cohorts resulted from the large volume of births that occurred from 1946 up through 1964. As these cohorts age over time, they tend to swell the sizes of the age categories they happen to occupy. By the year 2030, the size of the elderly population will be large because all of the baby boomers will have aged past their 65th birthday by then.

The age structure of the U.S. population can also change as a result of immigration. Typically, immigrants come here in their young adult ages and immediately increase the nation's population sizes of those ages. Past immigration can also affect the population sizes of older age groups. For example, the large immigration waves in the early twentieth century now constitute a part of America's elderly population.

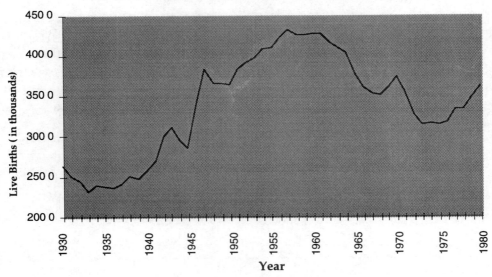

Live Births in the United States 1930-1980

KEY concepts

Age The age of the person in complete years according to the person filling out the census. Ages are usually grouped into 5 or 10 year age groups for analysis purposes.

Gender _Male_ or _female_.

Year Census years in which data are collected, in ten year intervals, or future years for projected populations.

Cohort The number of people born in a specific period usually in the same decade, sharing a common set of historical experiences. The cohorts examined here will be given names associated with approximate historical periods, as follows:

> _Roaring Twenties Cohort_ born 1916-1925
> _Depression Cohort_ born 1926-1935
> _World War II Cohort_ born 1936-1945
> _Early Baby Boom Cohort_ born 1946-1955
> _Late Baby Boom Cohort_ born 1956-1965
> _Generation X Cohort_ born after 1965

NOTE: In the datasets, there is no separate "cohort" variable. Instead, you can identify a cohort from an age group in a given year (See Figure A). For example, the 25-34 year old age group in 1990 belongs to the "Late Baby Boom Cohort".

State Refers to the 50 U.S. states and the District of Columbia (Washington D.C.)

Region Refers to groups of states (see Figure B) which are often used in studies of U.S. regional differences: _Northeast_, _Midwest_, _South_, and _West_.

City-Suburb-Nonmetropolitan Refers to the type of geographic area a person lives in. People who live in a metropolitan area either live in the central city or surrounding suburbs. Persons who live outside metropolitan areas are classed here as non-metropolitan (variable GEO3 in datasets).

OTHER concepts

Education (Topic two)   **Immigration Status** (Topic three)
Marital Status (Topic five)   **Race/Ethnicity** (Topic two)

The national population structure is also affected by mortality. In the United States, high levels of mortality do not occur in age groups under 65. This means that decreases in the population due to deaths do not have a great effect on the age structure of the under-65 population.

Of course, both cohort aging and immigration affect other aspects of the age structure of the United States. For example, cohorts born before 1945 do not have the same educational attainment as those born in the 1950s and 1960s. The race/ethnicity composition of the U.S. population also differs by age. This can be explained, in part, by the recent immigration from Latin America and Asia, which increases the Latino and Asian populations in the young adult age groups. The higher fertility of Latinos contributes to a larger number of Latinos in the child population.

Finally, although most of this chapter focuses on the nation as a whole, it is useful to look at the population structure of smaller geographic areas like census regions, states, and city-suburb breakdowns. The population structure of these smaller geographic areas is affected by migration within the United States as well as cohort aging, and immigration from abroad.

$\mathcal{A}.$ Cohorts and Changing Age Structure

A cohort refers to the number of people born in a specified period. The number of births that occurred between 1946 and 1964 was so large that the birth cohorts for all of these years have been given the blanket term "baby boom cohorts". Still, one can distinguish between "early baby boom cohorts" (born between 1946 and 1955) and "late baby boom cohorts" (born between 1956 and 1964). The names of other cohorts are either related to their size or the historical period in which they were born. We also show how the size of the birth cohort in the given period later affects the age structure of the population.

Exercise 1 The size of the population between the ages of 0-4 provides an indication of the recent fertility levels of the United States. Look at the number of persons 0-4 years old at successive 10-year intervals, from 1930 to 2000. Do these patterns suggest high levels of fertility during years the baby boom cohorts were born? *(Use the dataset POPSTRUC.DAT in the CENTrend directory)*

■ Create a line graph showing the number of children, ages 0-4, for each census year from 1930 through 2000. (Hint: The numbers in *POPSTRUC.DAT* are in 1000s)

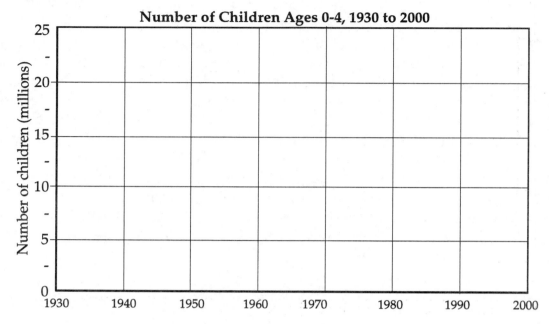

Number of Children Ages 0-4, 1930 to 2000

Exercise 2 Refer to figure A and notice that for any given year, each age group can be identified by its cohort (or birth period). For example, in 1990, the Early Baby Boom cohorts (born between 1946 and 1955) were 35-44. Determine the age group distribution of the 1990 U.S. population. What does this tell you about the relative size of each cohort? (*POPSTRUC.DAT*)

Figure A: Ages of Different Cohorts 1950-1990

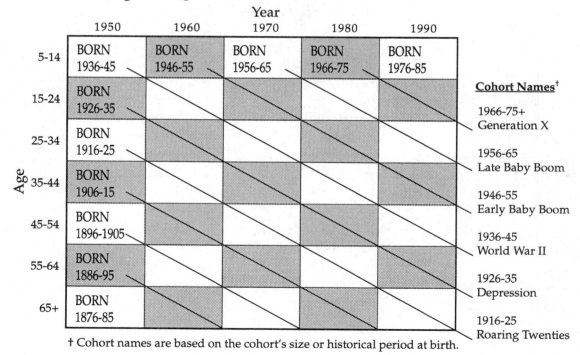

† Cohort names are based on the cohort's size or historical period at birth.

■ Create a bar chart showing the percentage of each age group in the 1990 U.S. population.

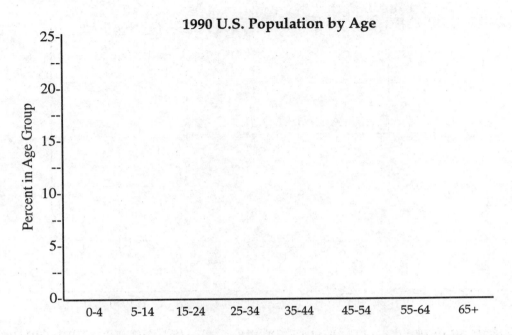

1990 U.S. Population by Age

Exercise 3 Now determine the age group distribution of the 1970 U.S. population. Which groups are the largest? Smallest? Are these results consistent with what you found for 1990? (*POPSTRUC.DAT*)

■ Create a bar chart showing the percentage of each age group in the 1970 U.S. population.

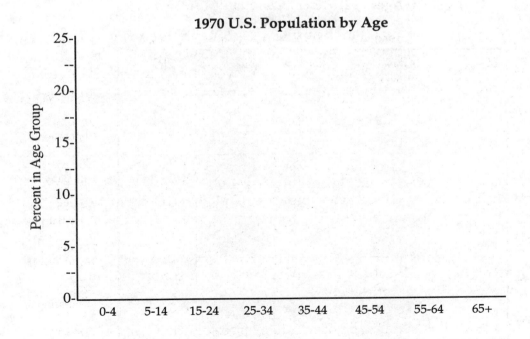

1970 U.S. Population by Age

Exercise 4 People begin to enter the labor force between the ages of 15 and 24. How did the aggregate size of this new labor force population age group change between 1950 and 1990? What was the impact of the early baby boom cohorts? Late baby boom cohorts? And the smaller "early generation X" cohorts? (*POPSTRUC.DAT*)

■ Create a line graph showing the number of persons ages 15-24 for each census year between 1950 and 1990.

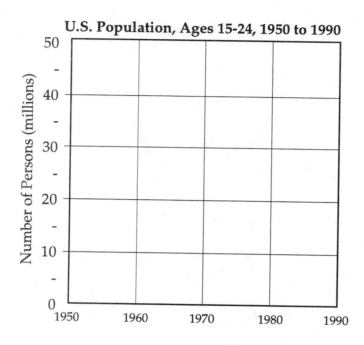

Exercise 5 Ages 25-34 are often the prime ages for entering the housing market. On your own, repeat exercise 4 for the 25-34 age group. What does this imply about change over time in the number of first-time home buyers? How would you interpret these changes in terms of different cohort sizes? (*POPSTRUC.DAT*)

$\mathcal{D}$**iscussion Questions**

1. How do fluctuating cohort sizes impact age-related societal institutions, like public school systems or nursing homes?

2. What effect will the retirement of the baby boom cohorts after the year 2010 have on the social security system? How is your cohort likely to fare from this program after retirement?

3. Do you think it is better to be born into a large cohort or a small cohort? Explain.

B. Women Live Longer

An almost equal number of boys and girls are born into a cohort, and the relative proportion of men and women does not change much over most of the life course. However, it is well-known that men have lower life expectancies than women and this affects the relative numbers of men and women in the oldest age categories.

Exercise 6 Calculate the percentage of the female population in each age group in 1990. For which ages does this rise substantially above 50 percent? (*POPUSA9.DAT*)

■ Create a bar chart with bars for each age group showing the percentage of females in 1990.

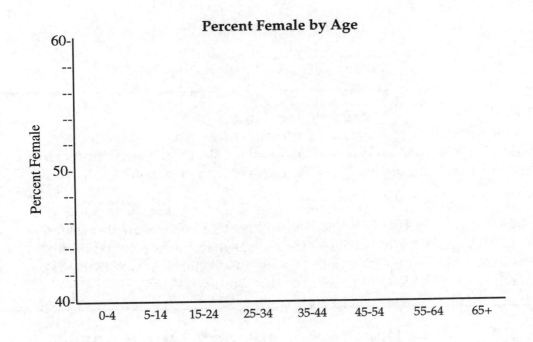

Percent Female by Age

Exercise 7 Has the percentage of females in the 65+ age group increased or decreased since 1950? What does this say about successive cohorts of elderly? (*POPSTRUC.DAT*)

■ Create a line graph showing, for the 65+ age group, the percentage of females for each census year between 1950 and 1990.

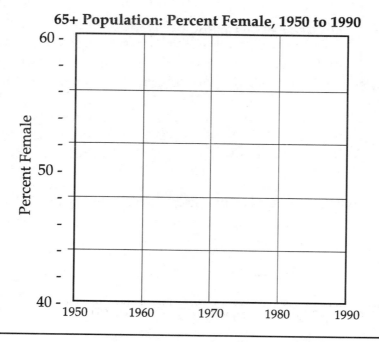

65+ Population: Percent Female, 1950 to 1990

Percent Female (y-axis: 40, 50, 60)

(x-axis: 1950, 1960, 1970, 1980, 1990)

---

*𝒟*iscussion **Questions**

1. What are some of the social and economic consequences of having a much larger female than male elderly population? What are the implications of this for the large baby boom cohorts as they enter their post-65 ages?

2. If women tend to marry men a few years older than they, how is the marriage market affected by a situation where large cohorts are immediately followed by small cohorts? (This was the case when the late baby boom cohorts were followed by the smaller "baby bust cohorts.") Does this make the marriage market "better" or "worse" for women? Why?

C. Baby Boomers, *Xers*, and Diplomas

Some social forecasters make the assumption that as today's younger cohorts age, they will take on the same social and economic attributes of older people in today's population. One social attribute that does not follow this assumption is educational attainment. This is because today's oldest age groups have lower education levels than the younger cohorts. During the 1950s and 1960s, nationwide improvements in public education had the greatest impact on the early baby boom cohorts and later cohorts.

Exercise 8 For the 1990 population, examine age differences in the percentage of people who graduated from college. Which age groups show the highest and lowest percentage of college graduates? What cohorts did they represent? *(EDUC5090.DAT)*

■ Create a bar chart with bars for each age group showing the percentage who have graduated from college.

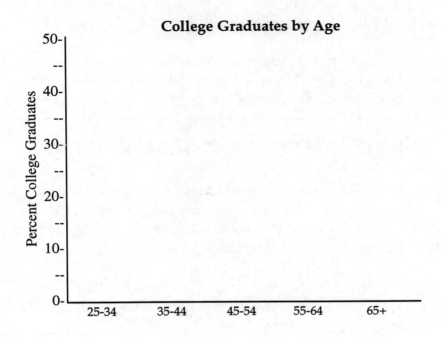

College Graduates by Age

Exercise 9 Contrast the educational attainment of 25-34 year olds in 1950 with those in 1990. Where do the main differences lie? (*EDUC5090.DAT*)

■ Create two pie charts, one for 1950 and one for 1990. In each chart, indicate the educational attainment distribution of 25-34 year olds.

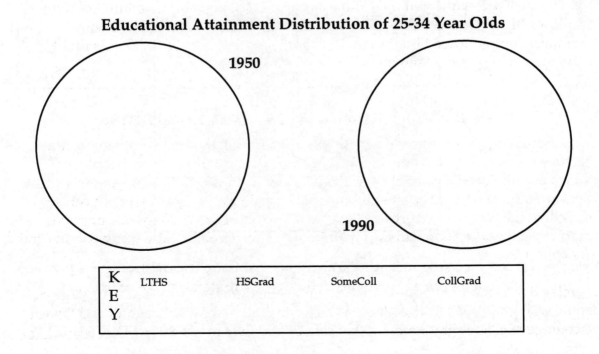

Educational Attainment Distribution of 25-34 Year Olds

1950

1990

| K E Y | LTHS | HSGrad | SomeColl | CollGrad |

$\mathcal{D}$iscussion Questions

1. What social and economic conditions do you think prompted the significant improvement in the U.S. educational system during the 1950s and 1960s?

2. Is a college education more valuable today than it was for young adults in the 1950s? If so, do you think that future generations will be even more likely to graduate from college than they are today?

$\mathcal{D}$. Cohort Differences in Marital Status

The 1950s were a period when couples married early and tended to stay together until "death do us part". This is less common today. Similar to the education comparisons above, we cannot look at the marital status experiences of our older population as a likely path to be taken by today's younger cohorts.

Exercise 10 Were young adults really more likely to be married in 1950 and 1960 than in 1990? For persons ages 25-34, calculate the percentage "currently married" in each census year from 1950 through 1990. *(MARR5090.DAT)*

■ Create a line chart showing the percentage of 25-34 year olds "currently married" in each census year from 1950 through 1990.

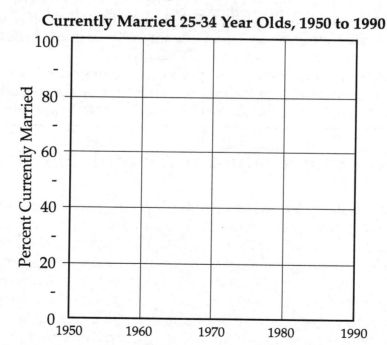

Currently Married 25-34 Year Olds, 1950 to 1990

Exercise 11 Now looking just at 1950 and 1990, compare the percentage of people married in each age group. Have the trends away from pervasive marriage occurred across the board? Are they more evident for particular age groups or cohorts? *(MARR5090.DAT)*

■ Create a bar chart with side by side bars for 1950 and 1990. For each age group, show the percent currently married.

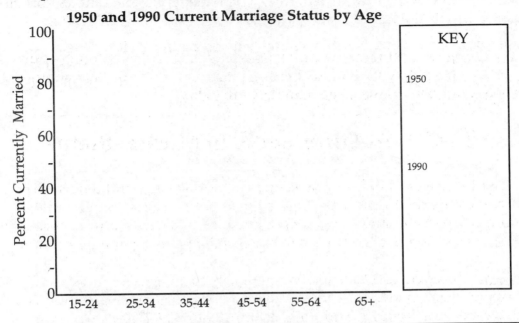

1950 and 1990 Current Marriage Status by Age

KEY

1950

1990

(y-axis: Percent Currently Married, 0–100; x-axis: 15-24, 25-34, 35-44, 45-54, 55-64, 65+)

𝒟iscussion Questions

1. Why do you think young adult cohorts since the mid-60s have been less inclined to marry and remain married for long periods of time? Do you think there may be a return to more stable and long-lasting marriages?

2. What does the overall aging of the population, along with lower life expectancy for men, imply for the marital status and living arrangements of older women?

𝓔. Recent Immigration and Population Structure

Immigration to the United States introduces another component of population change to the nation's age structure. This component has been especially important since the 1965 Immigration Act which led to both a rise in the number of immigrants and a change in the regions of origin. The majority of immigrants now come from either Latin America or Asia. Immigration to the United States is highly concentrated in particular regions, but it also affects the total population's race/ethnicity composition. Not all immigrants to the United States arrived since 1965, however. Some of our immigrant population arrived between 1880 and 1930 and these foreign-born citizens now comprise part of our nation's older population. A much smaller number of immigrants arrived between 1930 and 1965 as a result of restrictive immigration laws, the Great Depression, and disruptions due to World War II.

Exercise 12 Show, for each age group in 1990, the percentage of the foreign-born population who arrived before 1970 and since 1970. (Note: the 1965 Immigration Act was not implemented until 1968, so 1970 represents a reasonable cut-off year.) (_POPUSA9.DAT_)

■ Create a stacked bar chart with bars for each age group; stack by the percentage of native-born people, immigrants who came before 1970, and those who arrived between 1970 and 1990. (Hint: Combine foreign-born categories 1970-79 and 1980-90 to get 1970-90.)

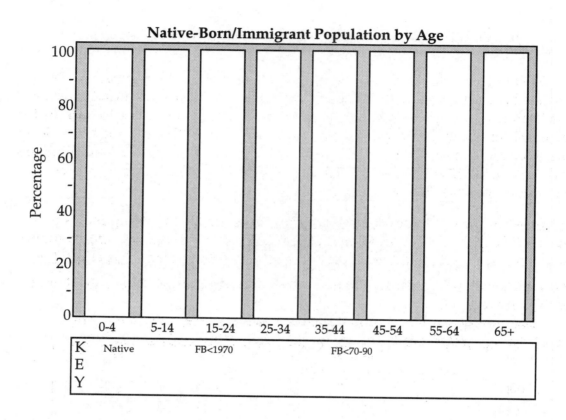

Exercise 13 On your own, compare the _numbers_ of foreign-born across age groups (rather than their percentage of the population). Again, look separately at those who arrived before 1970 and those who came between 1970 and 1990. What do these numbers indicate about the age structure of recent immigrants to the United States? (_POPUSA9.DAT_)

1. Consider the impact of post-1970 immigration on the nation's age structure. Does it help to increase the younger population at a time when the larger baby boom cohorts are growing older?

2. The percent of the 0-4 age group that is native-born is much higher than in any other age group. Why do you think this is the case?

$\mathcal{F}$. Immigration, Race/Ethnicity, and Age

The United States is comprised of many race and ethnic groups. Commonly used categories to identify these groups include blacks, Asians (including Pacific Islanders), American Indians (Eskimos and Aleuts), whites, and Latinos (or Hispanics)—see "Key Concepts" in Chapter 2. While all of these groups have long histories of residence in the U.S., Asians and Latinos have increased their numbers recently as a result of immigration. These two groups are most greatly represented among younger adult and child age groups.

Exercise 14 Contrast the race/ethnicity composition of the following four age groups in 1990: 5-14, 25-34, 45-54, and 65+. What differences do you find between the age groups? How would you account for these differences? (*POPUSA9.DAT*)

■ Create a stacked bar chart with bars for each age group (5-14, 25-34, 45-54, and 65+); stack by race/ethnicity.

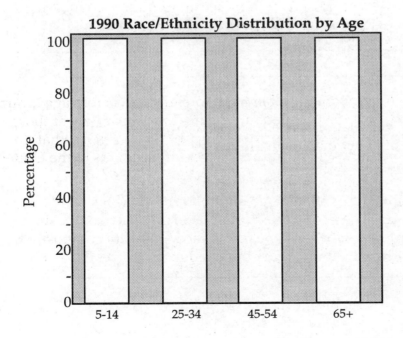

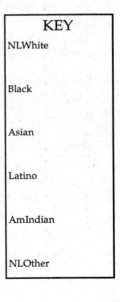

Exercise 15 To what extent do you think the foreign-born population contributes to the large Latino and Asian populations in the 25-34 age group? Examine this by comparing the 1990 race/ethnicity distribution of the foreign-born population of 25-34 year olds to the native-born population of the same age. *(POPUSA9.DAT)*

■ Create two pie charts, one for the combined foreign-born population, ages 25-34, and one for the native-born population of the same age. In each pie, show the race/ethnicity distribution.

1990 Race/Ethnicity Distribution of 25-34 Year Olds by Immigration Status

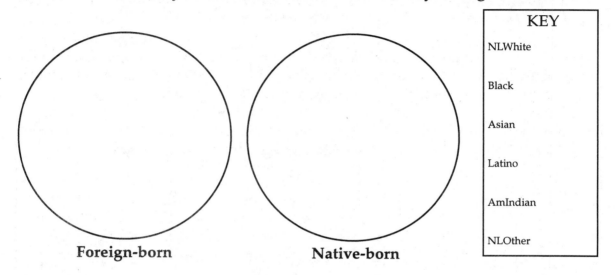

Foreign-born   **Native-born**

KEY

NLWhite

Black

Asian

Latino

AmIndian

NLOther

Exercise 16 On your own, compare the age distributions of whites, blacks, Latinos and Asians in 1990. To what extent do you think recent immigration may contribute to the differences in the age distributions of these groups? *(POPUSA9.DAT)*

𝒟iscussion Questions

1. In 1990, the elderly population had a lower percentage of racial and ethnic minorities than most of the younger age groups. Do you think this situation will change by the year 2010? Explain.

2. What role do you think the different fertility patterns of blacks, whites, Latinos, and Asians play in affecting the current race/ethnicity composition of the younger age groups? How will the future race/ethnicity composition of the U.S. be affected by mixed race and ethnic group marriages?

G. Geography

Until now we have concentrated on examining the entire U.S. population. It is often useful to focus on particular areas of the country, such as its 50 states (and the District of Columbia), or the four regions that the Census Bureau uses to categorize the country — Northeast, South, Midwest, and West (see Figure B). Another useful way to classify the population is by the type of area people live in. Most of the U.S. population lives in one of more than 300 metropolitan areas. Within these areas, they either live inside the central city or in the suburbs. Three categories can be used to classify the types of areas people live in: central city, suburbs, and non-metropolitan. In addition to these categories, there are many other ways to look at geographic distinctions across the national population down to small areas, like city blocks. The exercises below are based on examining states, regions, cities, suburbs, and non-metropolitan areas.

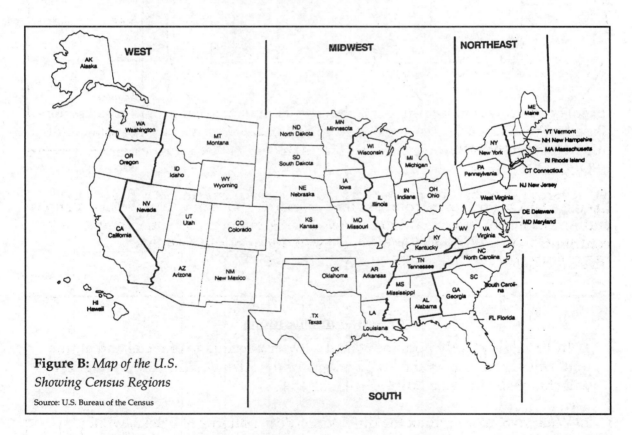

Figure B: *Map of the U.S. Showing Census Regions*

Source: U.S. Bureau of the Census

<u>**Exercise 17**</u> The age structures of regions often reflect their recent growth patterns. Growing regions tend to have higher percentages of young adults and children. This is because those who migrate within the U.S. tend to be younger than the general population. What can you tell about the recent growth patterns of the four census regions by examining their 1990 age distributions? (*POPGEO9.DAT*)

■ Create a stacked bar chart with bars for each census region in 1990 (Northeast, Midwest, South and West); stack by age groups.

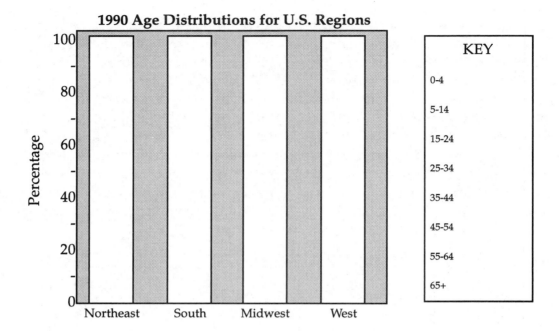

Exercise 18 Now compare the race/ethnicity distributions of the four census regions. Based on your results, which regions do you think have received a large number of recent immigrants? In which region is the percentage of blacks highest? (*POPGEO9.DAT*)

■ Create a stacked bar chart with bars for each census region in 1990; stack by the race/ethnic groups.

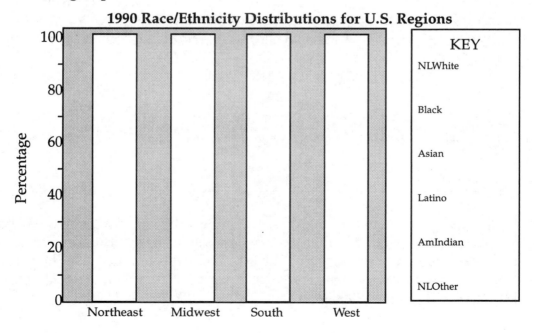

Exercise 19 Because recent immigration is heavily focused on particular states, these states will have higher percentages of Latino and Asian populations than most of the rest of the country. Compare the projected year 2010 race/ethnicity distributions of California, Texas, Florida, and New York with that of the total U.S. population. *(POPPROJ9.DAT)*

■ Create a stacked bar chart with bars for the total U.S. population, California, Texas, Florida and New York; stack by the 1990 race/ethnicity distribution.

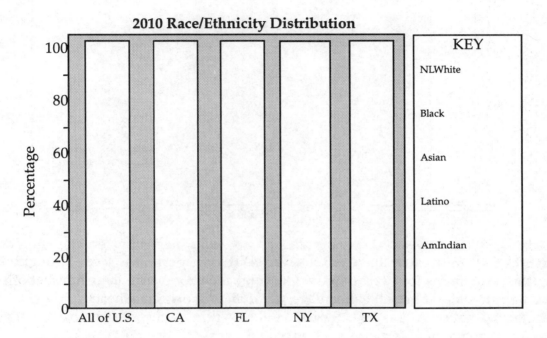

2010 Race/Ethnicity Distribution

Exercise 20 Central cities and suburbs are thought to be attractive to different age groups in the sense that "young singles" tend to want to live in the central city, while older adults with children tend to want to live in the suburbs. Find out if this is true by examining the percentage of children (ages under 15) and the percentage of young adults (ages 15-24) in each type of area. *(POPGEO9.DAT)*

■ Create a bar chart with side by side bars for central cities and suburbs. For each age group (under 15 and 15-24), show the percent living in central cities and suburbs.

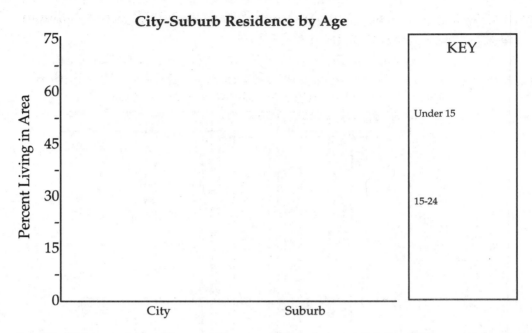

City-Suburb Residence by Age

Exercise 21 Non-metropolitan areas are said to have a higher percentage of elderly residents. This is because some of these areas attract elderly due to their amenities and other attractions for new retirees. Other non-metropolitan areas have a high percentage of elderly because they have lost a disproportionate share of their younger populations. For the U.S. as a whole, do non-metropolitan areas have a higher percentage of elderly than metropolitan areas? (*POPGEO9.DAT*)

■ Create two pie charts, one for the metropolitan areas and one for non-metropolitan areas. In each pie, show the 1990 age distribution. (Hint: Combine categories City and Suburb.)

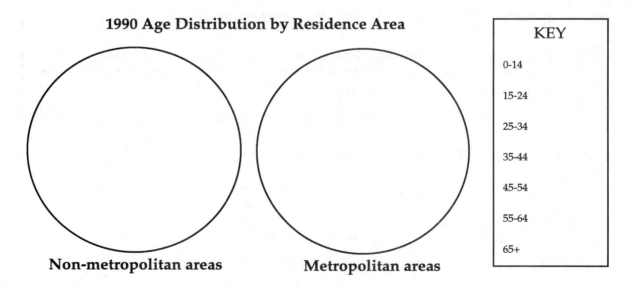

1990 Age Distribution by Residence Area

Exercise 22 Examine the race/ethnicity compositions of central cities, suburbs and non-metropolitan areas. Are they as different as you had expected them to be? If not, why not? *(POPGEO9.DAT)*

■ Create a stacked bar chart, with bars for city, suburb, and non-metropolitan areas; stack by 1990 race/ethnicity distribution.

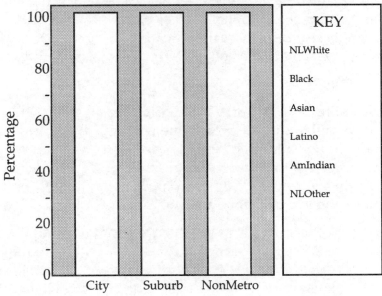

Race/Ethnicity Distributions by Geographic Area

KEY

NLWhite

Black

Asian

Latino

AmIndian

NLOther

𝒟iscussion Questions

1. Why do national patterns of age structure and race/ethnicity composition not necessarily characterize individual states, regions or cities?

2. Do you agree with this statement: It is easier to predict the future age structure of the U.S. population than it is to project the future age distribution of an individual state. Why or why not?

THINK *tank*

1. We have mentioned that the patterns of marriage, divorce, and family life followed by most of the pre-baby boom cohorts have not been followed by the early or late baby boomers. Similarly, the educational levels of these early cohorts were more limited than for people born since 1945. In light of these trends, and what the data already show, speculate on the future marital patterns, childbearing, and occupational choices of your own cohort. How might changing economic and social realities lead to different choices for your cohort than the pre-baby boom cohorts? The baby boom cohorts? Elaborate.

2. California, Texas, Illinois, New York, and Florida are all states with diverse race/ethnic populations, but which one will have the largest increase in the percentage of Latino children by 2010? Based on 1990 data, would you expect each Latino subgroup to show the same amount of increase in each state? Are some Latino subgroups likely to be more influential in some states than others? Explain and support your answers.

RACE AND ETHNIC INEQUALITY
topic two

The population of the United States is becoming more racially and ethnically diverse. This increasing diversity is an important issue because it is changing the cultural, political, and economic landscape of American life. Our schools, workplaces, legislatures, and national character are constantly being shaped by this growing diversity. Consequently, race/ethnicity issues concern the entire nation, not just the members of minority groups.

Before examining the similarities and discrepancies between different race/ethnic groups, it is necessary to gain a historical perspective. Various events in America's past have contributed to the current state of race/ethnic inequality in the United States. The legal oppression of African-American slaves and the implementation of Jim Crow laws set the stage for this inequality. However, in the mid-twentieth century, this unjust and discriminatory treatment began to be challenged and vigorously opposed by many. In this context, the Civil Rights Movement began.

The Supreme Court decision in *Brown v. Board of Education* served as one of the starting points of this movement. The ruling determined that separate facilities for blacks were inherently unequal and that segregation was no longer constitutional. Although this was a significant victory for the supporters of the Civil Rights Movement, the battle had not yet been won. Another triumph came with the passage of the Civil Rights Act of 1964. This act determined that the nation would no longer recognize legal distinctions based upon race, color, creed, or national origin in the workplace. Any employer who did not honor this newly established equality would be in violation of United States law.

Today, the movement's struggle is embodied in the creation and implementation of Affirmative Action programs, which apply to several ethnic groups. These programs have increased minorities' access to colleges and workplaces by enforcing specific and compulsory admittance and hiring guidelines. Although some people

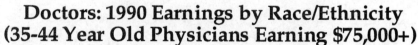

Doctors: 1990 Earnings by Race/Ethnicity
(35-44 Year Old Physicians Earning $75,000+)

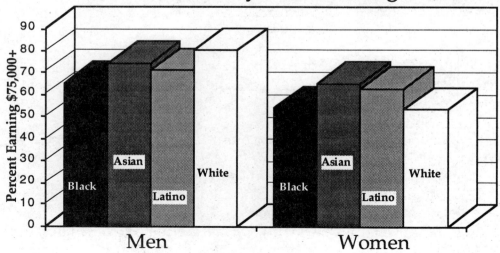

KEY concepts

Race/Ethnicity Identifies the major ethnic groups, combining the race and Hispanic-origin classifications used by the Census Bureau based on a person's self-identification.

Non-Latino White (NLWhite) all persons who indicated their race as white and not of Latino origin.

Black all persons who indicated their race as black.

Latino (Hispanic Origin) persons of white or "other" races who identified themselves as Mexican, Puerto Rican, Cuban, or Other Spanish/Hispanic. This category can refer to ancestry, nationality group, lineage, or country of birth of the person's parents or ancestors before their arrival in the U.S..

Asian (Asian or Pacific Islander) includes all persons who indicated their race or ethnicity as Chinese, Filipino, Japanese, Asian Indian, Korean, Vietnamese, Cambodian, Hmong, Laotian, Thai, or Other Asian. Also includes persons who indicated their race as Hawaiian, Samoan, Guamanian, or other Pacific Islander.

Amercian Indian (American Indian, Eskimo, or Aleut) all persons who classified themselves as American Indian, Eskimo, or Aleut.

Other (NLOther) includes persons who indicated *other* in the race classification and are not of Latino origin. This category also includes people who identified themselves as interracial, multiracial, multiethnic, mixed, or Wesort.

NOTE: The RACE and RACELAT variables in the datasets often combine some of these categories. (For example, RACELAT5 combines the category of American Indians with Other.)

Latino Groups Hispanic-Origin persons of any race can be specified based on their self-identification for the census according to the following categories: *Mexican*, *Puerto Rican*, *Cuban*, *Central American*, *South American*, and *Other* (variable LATINO6 in datasets).

Asian Groups Persons indicating specific race or ethnicity as: *Chinese*, *Japanese*, *Filipino*, *Korean*, *Indian*, *Vietnamese*, or *Other Asian* (variable ASIAN7 in datasets).

Education The highest level of school completed or the highest degree received.

<9 Years persons who have completed less than 9 years of schooling

9-12 Years persons who have completed 9-12 years of schooling but have not graduated from high school

High School Graduate persons who have graduated from high school

Some College persons who have completed some years of college or attained an Associate Degree

College Graduate persons who have graduated from college

Master's Degree persons who have completed an MA, MS, Med, MSW, MBA, or other similar degree

Ph.D. or Professional Degree persons who have completed a doctorate level degree (Ph.D, Ed.D) or professional school degree (MD, DDS, DDM, LLB, JD)

NOTE: The EDUC variables in the datasets sometimes combine these categories. For example, the category "CollGrad" in variable EDUC4 refers to persons who have a college degree or more education.

Occupation The classification system for this category has changed over the years. It includes all employed workers, and, in its simplest form, divides them into the following:

Top White Collar professional writers, executives, administrators, and managers

Other White Collar administrative support, clerical and sales workers, technicians, and related support

Service private household, protective service, and other service workers

Top Blue Collar "skilled blue collar" jobs such as precision production, craft, and repair workers.

Other Blue Collar workers in less skilled blue collar jobs

Farm workers in farm, forestry, and fishery occupations

NOTE: The OCCUP variables in the datasets sometimes combine these categories. For example, the variable OCCUP5 combines the farm category with Other Blue Collar. The variable OCCUP4 combines Top Blue Collar, Other Blue Collar, and Farm with Blue Collar.

Earnings Money a person makes from working, as wages, salary, or a form of self-employment, expressed as an annual amount.

may feel Affirmative Action programs are unjust, few can deny that these programs have done a great deal for the diversification of America's workplaces and educational institutions.

While keeping these historical events in mind, you will look at the similarities and discrepancies between different race/ethnic groups in terms of educational attainment, occupations and earnings. Over time, all race/ethnic groups have experienced increased education levels, more occupational choices, and higher earnings. However, the rate of these gains varies between race/ethnic groups. After seeing the gaps between race/ethnic groups, you will consider why these discrepancies exist. As you work through the following exercises, consider whether we have made much progress towards race/ethnic equality since the Civil Rights Act of 1964. What evidence of racial discrimination still remains in society today?

𝒜. Education and Race/Ethnicity

Over the last few decades, the educational attainment of all race/ethnic groups in the United States has increased steadily. Several factors have led to this increase. First, the mandate for compulsory education for children under the age of sixteen contributed to an increase in the percentage of high school graduates. The court ruling in *Brown v. The Board of Education* also served as an influential force by increasing access to public education for black children. Since the 1950s, federal education initiatives have increased school opportunities for new cohorts of young adults. Affirmative action policies helped to provide higher education opportunities that were previously unavailable to many minority students. In addition to legal mandates, changes in the United States economy and industrial structure have influenced changes in educational needs and levels. As our society moves from a manufacturing-based industrial economy into an age of technology and information, higher education has become the key to entry into stable, well-paying jobs.

Still, educational attainment has not increased for all race/ethnic groups at an equal rate. In this section, you will examine trends in educational attainment since 1950 and current race/ethnicity gaps. As you work through the exercises, consider why the educational attainment of race/ethnic groups has not increased at an equal rate.

Exercise 1 Using data from 1950 to 1990, examine changes in the percentage of blacks and nonblacks, ages 25-34, with a high school education or more. Describe the overall trends as well as how differences between the two groups have changed over time. Why do you think it is useful to focus on the 25-34 year old age group for this trend?

■ Create a line graph with two lines, one for blacks and one for nonblacks, ages 25-34. For each year, indicate the percentage of high school graduates (or more) in each group. (Hint: Combine high school graduates, those with some college and college graduates.) *(EDUC5090.DAT)*

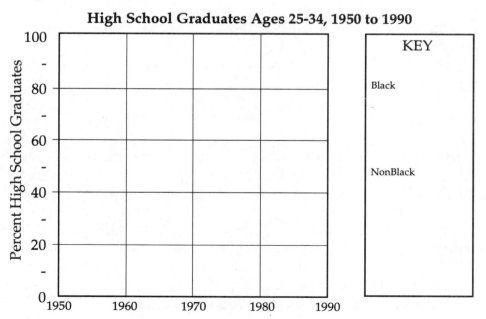

High School Graduates Ages 25-34, 1950 to 1990

Exercise2 Using data from 1950 to 1990, examine changes in the percentage of blacks and nonblacks, ages 25-34, who have graduated from college. How do the overall trends and the differences between the two groups compare to your findings for high school graduates? Give possible explanations for the racial gaps you may find. (*EDUC5090.DAT*)

■ Create a line graph with two lines, one for blacks and one for nonblacks, ages 25-34. For each year, indicate the percentage of college graduates in each group.

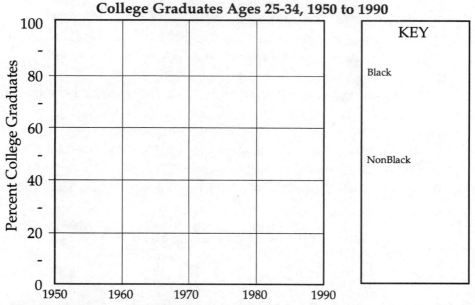

Exercise 3 Focusing on 1990, look at the educational attainment of people ages 25-34 in each race/ethnic group. Describe the differences between groups and give possible explanations. (*EDUCIMM9.DAT*)

■ Create a stacked bar chart showing the educational attainment distribution within each race/ethnic group; stack by education level.

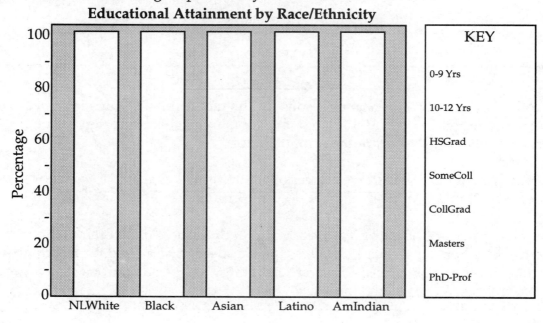

Exercise 4 On your own, repeat exercise 3 for people ages 45-54. Are the race/ethnicity gaps in education wider among the older cohorts? Offer possible explanations for your findings. *(EDUCIMM9.DAT)*

Exercise 5 Using 1990 data, look at the educational attainment of 25-34 year olds in each specific Asian group. Describe the differences between the groups. Overall, how do Asians' education levels compare to the educational attainment of other race/ethnic groups? *(EDUASN9.DAT)*

■ On your own, create a stacked bar chart showing the educational attainment distribution within each specific Asian group; stack by education level.

Exercise 6 On your own, repeat the previous exercise for specific Latino groups. *(EDULAT9.DAT)*

Exercise 7 Focusing on people ages 25-34, describe the race/ethnicity distribution of people holding Ph.D's and other professional degrees in 1990. Compare the representation of each race/ethnic group to whites and other groups. *(EDUCIMM9.DAT)*

■ Create a pie chart with divisions for each race/ethnicity.

Ph.D and Professional Degree Holders

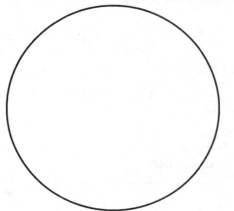

KEY

NLWhite

Black

Asian

Latino

AmIndian

NLOther

---

*D*iscussion Questions

1. What do you think are some reasons for the race/ethnicity gaps in educational attainment? Can you think of any factors which have affected certain race/ethnic groups' educational opportunities?

2. How might you explain the educational attainment changes you saw over time, overall and within specific race/ethnic groups?

3. Consider your findings regarding trends in the educational attainment of nonblacks and blacks since 1950. Does there appear to be a need for Affirmative Action, or other programs designed to help improve the educational opportunities for blacks? Why or why not?

$\mathcal{B}$. Occupation and Race/Ethnicity

The past decade has witnessed great changes in the race/ethnicity composition of occupations in the U.S. labor force. Many occupations traditionally held by white males are gradually becoming more representative of the entire U.S. population. Minorities are filling more managerial and professional positions. In addition, the increased levels of education have improved the overall occupational status of minorities.

However, minorities are still faced with barriers to advancement. They are often clustered in the lower-status occupations and many continue to be discriminated against in hiring and promotions. Consequently, as doors seem to be opening for minorities, they often encounter closed doors farther along the career path. Many corporations have "glass ceilings" which prevent minorities from attaining executive positions.

In the following exercises, you will look at the race/ethnicity composition of several occupational categories and determine how this distribution has changed since 1950. You will also look at the 1990 race/ethnicity distribution of doctors and lawyers in an effort to gauge how many minorities have prestigious occupations.

Exercise 8 Using data from 1950 to 1990, look at the percentage of the population in each occupational category. What kind of trends do you see in the occupational distribution of men and women? (_EDOC5090.DAT_)

■ Create two stacked bar charts, one for men and one for women. In each chart, for each year, stack by occupational categories.

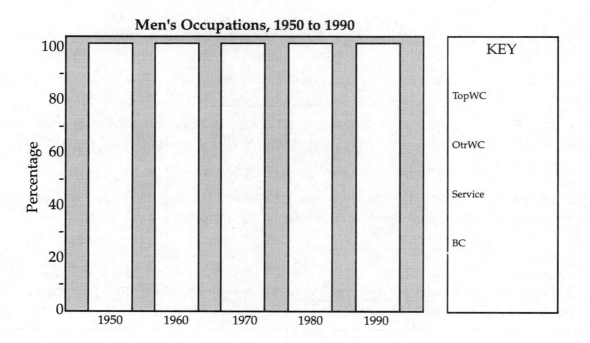

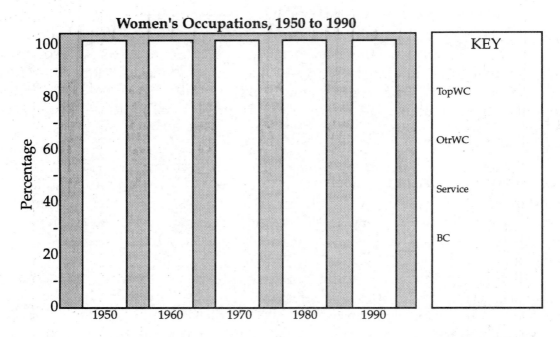

Women's Occupations, 1950 to 1990

KEY

TopWC

OtrWC

Service

BC

Exercise 9 Focusing on 35-44 year olds, show the occupational distribution of black and nonblack men from 1950 to 1990. Compare the distribution of these two groups to each other and to the overall occupation trends you saw in exercise 8. Why is it useful to look only at 35-44 year olds when making occupation comparisons? (*EDOC5090.DAT*)

■ Create four line graphs, one for each occupational category. Draw two lines in each chart, one for black men and one for nonblack men, ages 35-44; for each year, indicate the percentages employed in the specified occupational category.

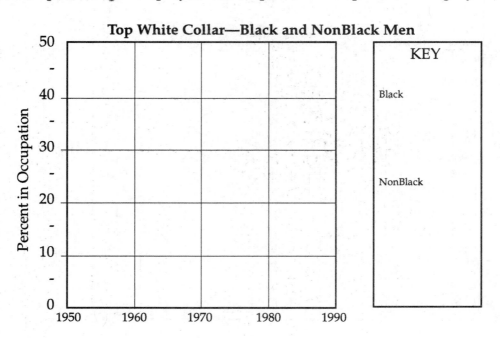

Top White Collar—Black and NonBlack Men

KEY

Black

NonBlack

Other White Collar—Black and NonBlack Men

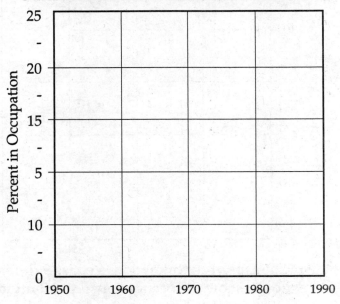

Service—Black and NonBlack Men

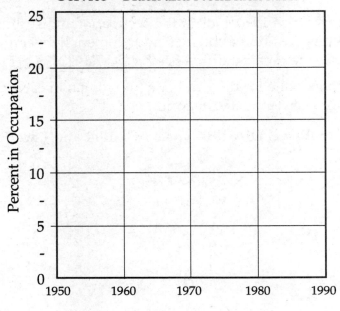

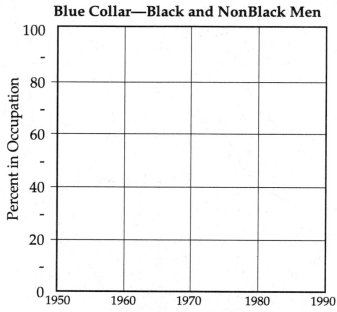

Blue Collar—Black and NonBlack Men

Exercise 10 Are the racial gaps you explored in exercise 9 different for women? Focusing on 35-44 year olds, show the occupational distribution of black and nonblack women from 1950 to 1990. Compare the distribution of these two groups to each other and to the trends you plotted in exercise 9. Why is it useful to look at men and women's occupational trends separately? *(EDOC5090.DAT)*

■ On your own, create four line graphs, one for each occupational category. Draw two lines in each chart, one for black women and one for nonblack women, ages 35-44; for each year, indicate the percentages employed in the specified occupational category.

Exercise 11 Focusing on 1990, look at the occupational distribution of 35-44 year old black, white, Asian, and Latino men and women. Between which race/ethnic groups do you see the greatest differences? *(OCCUPTN9.DAT)*

■ Create 8 pie charts, one for men and one for women in each race/ethnic group, with divisions for occupational categories.

Occupational Distributions by Gender and Race/Ethnicity

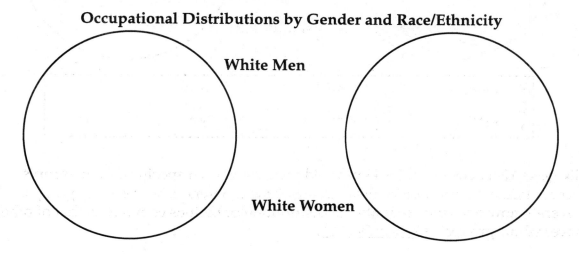

White Men

White Women

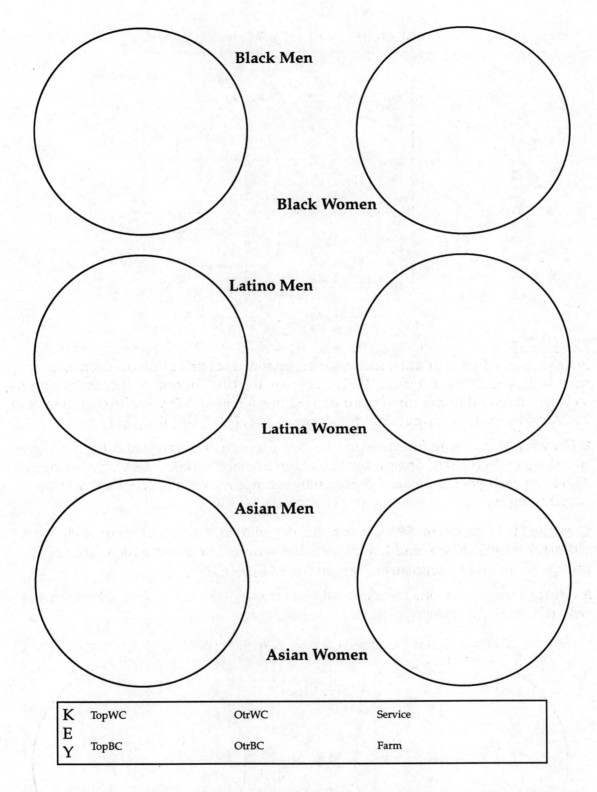

Black Men

Black Women

Latino Men

Latina Women

Asian Men

Asian Women

K E Y	TopWC	OtrWC	Service
	TopBC	OtrBC	Farm

Exercise 12 Focusing on 35-44 year old men, show each specific Latino group's occupational distribution in 1990. Describe the differences between the groups. Overall, how does the occupational distribution of Latinos compare to that of other race/ethnic groups? (*OCCLAT9.DAT*)

■ Create a stacked bar chart showing the occupational distribution within each specific Latino group; stack by occupational category.

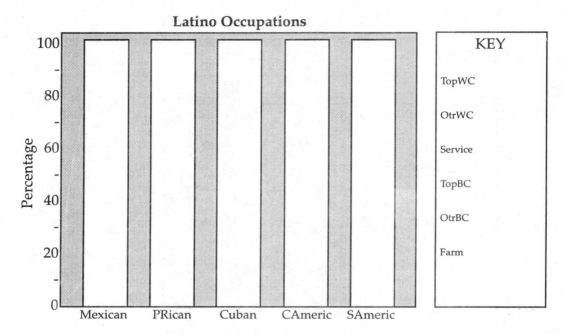

Latino Occupations

Exercise 13 On your own, repeat the previous exercise for each specific Asian group. (*OCCASN9.DAT*)

Exercise 14 Using 1990 data, show the occupational distributions of 35-44 year old men and women with different educational backgrounds. What is the apparent relationship between education and occupation? (*OCIM9-35.DAT*)

■ Create a stacked bar chart with side by side bars for men and women; for each educational category, stack by the occupational distribution.

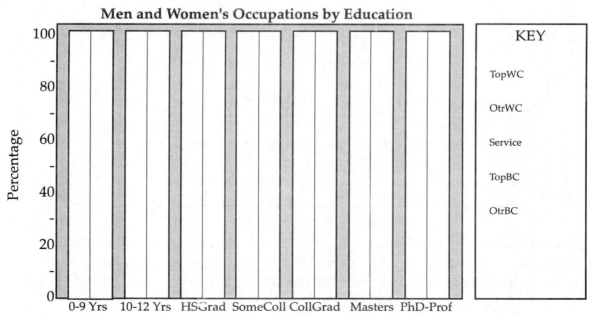

Men and Women's Occupations by Education

Exercise 15 Focusing on male college graduates ages 35-44 in 1990, examine the occupational distribution within each race/ethnic group. Describe your findings.

■ Create a stacked bar chart showing the occupational distribution within each race/ethnic group; stack by occupation. (Hint: Combine college graduates, M.A.'s, and Ph.D's.) *(OCIM9-35.DAT)*

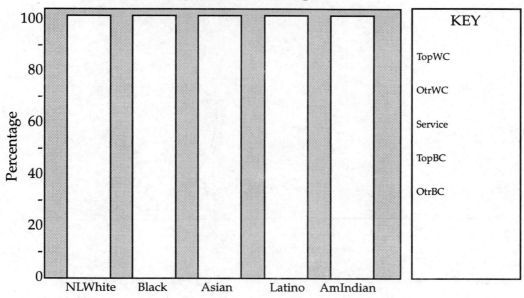

Male College Graduate Occupations

Exercise 16 Focusing on female college graduates ages 35-44 in 1990, examine the occupational distribution within each race/ethnic group. Describe your findings and how they differ from the results of the previous exercise.

■ Create a stacked bar chart showing the occupational distribution within each race/ethnic group; stack by occupation. *(OCIM9-35.DAT)*

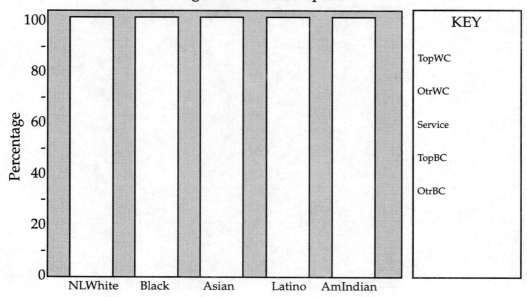

Female College Graduate Occupations

Exercise 17 Compare the race/ethnicity composition of doctors ages 25-34 to those ages 55-64. Discuss the representation of each race/ethnic group in this occupational category. Do you notice any differences between the two age groups? *(DOCTORS9.DAT)*

■ Create two pie charts, one for each age group, with divisions for each race/ethnic group.

Race/Ethnicity of Doctors

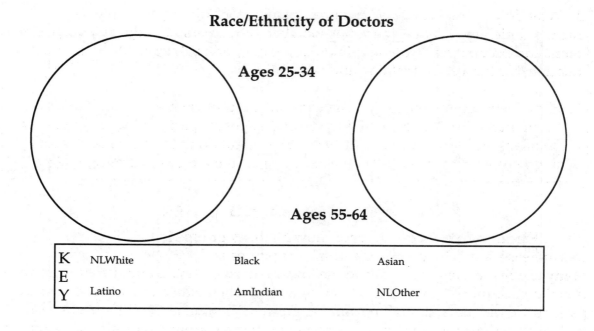

Exercise 18 Focusing on doctors ages 25-34, compare the race/ethnicity composition of male doctors to that of female doctors. How does controlling for gender affect the results from exercise 17? *(DOCTORS9.DAT)*

■ Create two pie charts, one for each gender, with divisions for each race/ethnic group.

Race/Ethnicity Distribution of Doctors by Gender

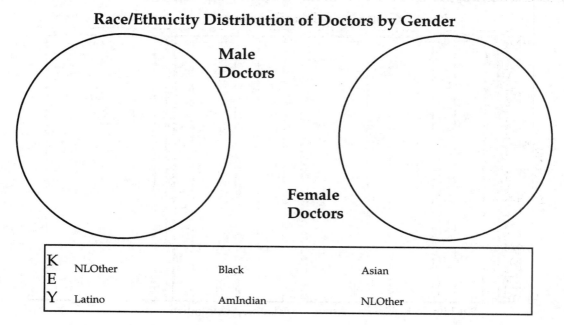

Exercise 19 On your own, repeat the previous two exercises for male and female lawyers ages 25-34 and 55-64. Describe the differences between age groups and genders. (*LAWYERS9.DAT*)

---

*D*iscussion Questions

1. Why do you think that occupational distributions differ between genders and races/ethnicities? What is the connection between a person's educational attainment and occupation? Do you think that gender and race/ethnicity affect the relationship between education and occupation?

2. How do you think the historical events of this century have affected the overall occupational distribution? Think about changes in the percentage of people in blue collar jobs between 1950 and 1990. What trend do you see in this category? Why do you think that there have been changes in the types of jobs people have?

C. Earnings Inequalities

Earnings rose moderately for most minority groups in the 1980s. Despite this increase, a gap between the earnings of whites and minorities still exists. Many argue that this gap is due to the lower education levels and different occupational distribution of minorities. However, even when education levels and occupations are the same, differences among groups still remain. In this section, you will examine earnings differences among race/ethnic groups with similar levels of education and occupations.

Exercise 20 Using 1990 data, illustrate the earnings distributions of men, ages 35-44 in each race/ethnic group. Describe any significant differences. (*EARN9.DAT*)

■ Create a stacked bar chart with bars for each race/ethnic group; stack by earnings.

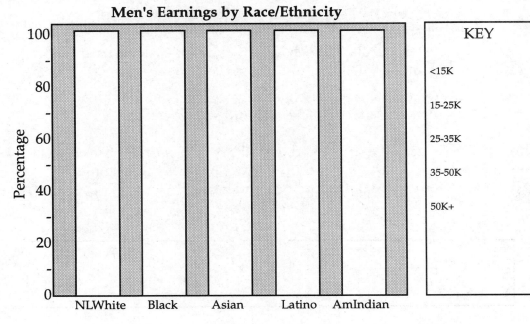

Men's Earnings by Race/Ethnicity

Exercise 21 How do women's earnings differ from men's earnings? Are there similar differences between race/ethnic groups? Using 1990 data, illustrate the earnings distributions of women ages 35-44. Describe any significant differences. (*EARN9.DAT*)

■ Create a stacked bar chart with bars for each race/ethnic group; stack by earnings.

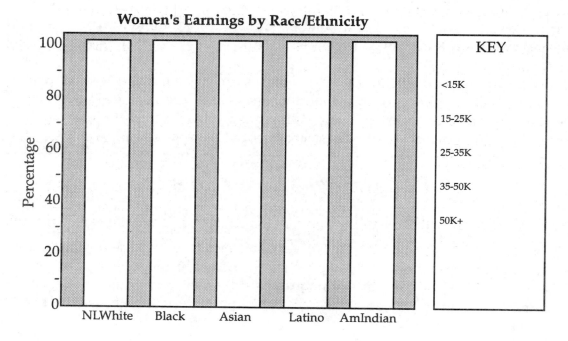

Exercise 22 Using 1990 data, examine the earnings distribution in each occupational category. What is the relationship between earnings and occupation? After determining the relationship between occupation and earnings, look specifically at the earnings of men, ages 35-44, in top white collar occupations, for each race/ethnic group. Describe the differences you find. (*WORK9-35.DAT*)

■ On your own, create a stacked bar chart with bars for each occupational category; stack by earnings.

■ On your own, create a stacked bar chart only for men in top white collar positions. Draw bars for each race/ethnic group and stack by earnings.

Exercise 23 On your own, explore and describe earnings differences between 35-44 year old men of specific Asian groups in 1990. How does the earnings distribution vary? Offer possible explanations for your findings. (*EARNASN9.DAT*)

Exercise 24 On your own, repeat the previous exercise for specific Latino groups in 1990. (*EARNLAT9.DAT*)

Exercise 25 On your own, focus on full-time, year round male workers ages 35-44. Determine if race/ethnicity gaps in earnings are mostly due to: a) race/ethnicity differences in educational attainment or b) race/ethnicity differences in occupation. (*WORK9-35.DAT*)

Exercise 26 On your own, explore and describe earnings differences between 35-44 year old women of specific Asian groups in 1990. How does the earnings distribution vary? Offer possible explanations for your findings. (*EARNASN9.DAT*)

Exercise 27 On your own, repeat the previous exercise for specific Latino groups in 1990. (*EARNLAT9.DAT*)

Exercise 28 On your own, focus on full-time, year round female workers ages 35-44. Determine if race/ethnicity gaps in earnings are mostly due to: a) race/ethnicity differences in educational attainment or b) race/ethnicity differences in occupation. (*WORK9-35.DAT*)

Exercise 29 Show the earnings distribution of all doctors in 1990. (*DOCTORS9.DAT*)

■ Create a bar chart with bars indicating the percentage of doctors in each earnings category. (Hint: Use the Marginals command.)

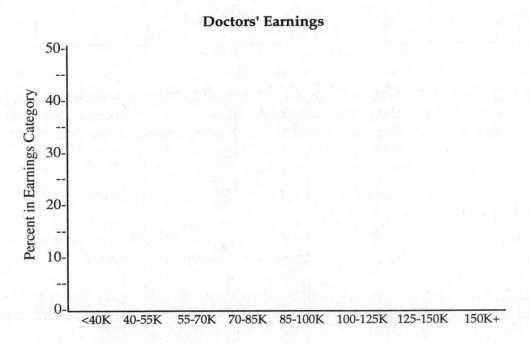

Doctors' Earnings

Exercise 30 Focusing on doctors ages 25-34 and 55-64, show the earnings distribution by race. What earnings differences do you see between race/ethnic groups? Age groups? (*DOCTORS9.DAT*)

■ Create two stacked bar charts, one for each age group. Draw bars for each race/ethnic group; stack by earnings.

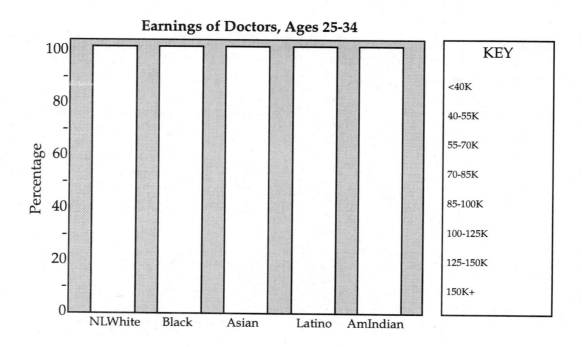

Earnings of Doctors, Ages 25-34

KEY

<40K

40-55K

55-70K

70-85K

85-100K

100-125K

125-150K

150K+

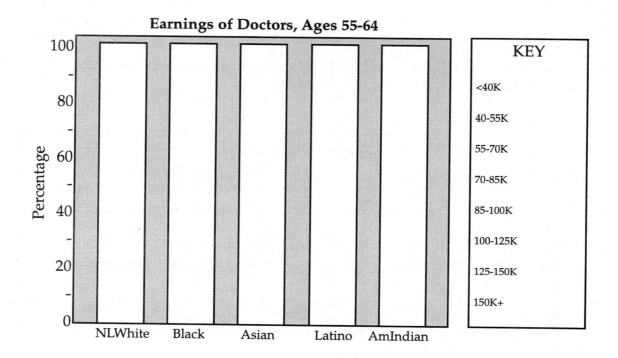

Earnings of Doctors, Ages 55-64

KEY

<40K

40-55K

55-70K

70-85K

85-100K

100-125K

125-150K

150K+

Exercise 31 Show the earnings distribution of all lawyers in 1990. (*LAWYERS9.DAT*)

■ Create a bar chart with bars indicating the percentage of lawyers in each earnings category. (Hint: Use the Marginals command.)

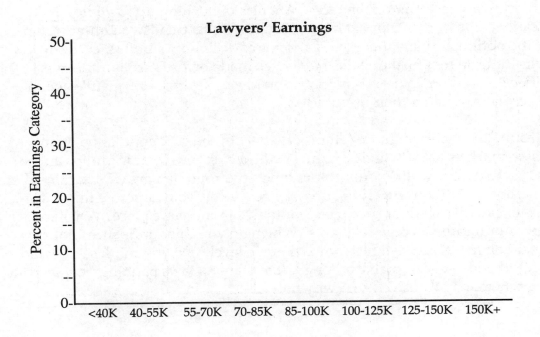

Lawyers' Earnings

Exercise 32 Focusing on lawyers ages 25-34 and 55-64, show their earnings distribution by race. What earnings differences do you see between race/ethnic groups? Age groups? (*LAWYERS9.DAT*)

■ On your own create two stacked bar charts, one for each age group. Draw bars for each race/ethnic group; stack by earnings.

𝒟iscussion Questions

1. What earnings differences exist between race/ethnic groups? How does educational attainment tend to affect different race/ethnic groups' earnings? Do all race/ethnic groups seem to experience similar "returns for education"?

2. How might you explain some of the differences discussed in the previous question? Why do you think these differences exist? Why have they persisted over time?

THINK *tank*

1. Some social critics argue that over the last few decades there has been a steady convergence among the races. Are blacks catching up to whites at all education, employment, and earnings levels? Do the trends you observe support the optimistic view that race is a factor of declining significance, or do your results indicate that further gains have to be made before such optimism is justified? Based on your examination, would you say that affirmative action policies have outlived their usefulness?

2. Despite the increasing race/ethnic diversity of the U.S. population, many housing markets are still highly segregated. Which race/ethnic groups appear most likely to live in cities, suburbs, and non-metropolitan areas? Examine these race/ethnic conditions by economic and age levels. Do the more affluent segments of each race/ethnic group appear to live in the same place? What about the poor or middle income segments? Which do you think influences housing segregation more, race/ethnicity or economic level? Do you think segregated neighborhoods are acceptable as long as there are no laws banning people from living where they want? Why or why not?

IMMIGRANT ASSIMILATION
topic three

Immigration has played an important part in the changing of America's economic, political and social landscape. Over its history, factors such as source regions, reasons for leaving and regions of settlement have changed dramatically, but immigration has remained a pertinent issue in the United States.

After the great influx of immigrants in the 1880s, immigration steadily declined and hit significantly low levels during the mid 20th century. However, with the 1965 Immigration Act and the easement of restrictive barriers, immigration was once again on the rise. During earlier periods, the United States saw most of its

Foreign Born Early Arrivals
(Percent of Group's Immigrant Population Arriving before 1970)

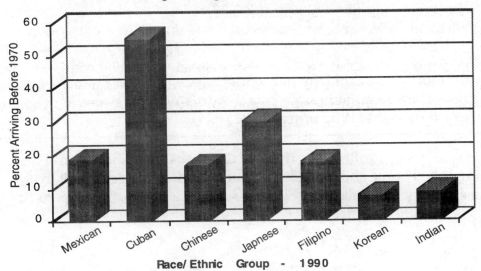

newcomers arriving from Eastern and Southern Europe and Canada. In recent decades, immigrants have been primarily from Asia, Mexico and other Latin American countries. This most recent trend is known as the "new immigration".

In any study of immigration, educational attainment, occupation and earnings must be considered. Social scientists often study the educational attainment of immigrants and relate it to their occupational attainment and earnings. It is helpful, as well as interesting, to see what types of jobs immigrants hold, for occupation is an indicator of both social and economic mobility. By looking at earnings, researchers can gauge the ease or difficulty immigrants have in reaching economic parity with native-born Americans.

In the following exercises, these factors, as well as others, will serve as your guide to studying immigration. You will look at who immigrates, where they settle and what they do once they are here. You will compare the experiences of recent immigrants with longer term residents. Keeping these comparisons in mind, you will consider to what degree immigrants are assimilating socially and economically, and to what degree they have maintained their own cultural heritage.

KEY concepts

Immigration Status U.S. residents can be classified as either native-born or foreign-born. The native-born population consists of persons born in the United States, including Puerto Rico and other outlying territories, as well as persons born abroad with at least one American parent. The foreign-born population includes all others born outside the United States. The foreign-born can be further classified by their year of entry into the U.S. The following categories are used here (dataset variable IMMIG4) to classify the 1990 U.S. population:

Native-born
Foreign-born, entered before 1970
Foreign-born, entered between 1970-79
Foreign-born, entered between 1980-89

NOTE: Dataset variables occasionally combine some of these categories.

Origin Country Foreign-born residents are classified by their country of birth. The following categories are used here (dataset variable ORIGIN6) to indicate the country or region of birth: Mexico, OtrLatAm(other Latin America), Asia, Africa, and NAEurOcn (North America, Europe, and Oceania).

English Language Proficiency The 1990 Census asked respondents if they spoke only English at home and if not, to indicate their own assessment of their English language ability. The following categories summarize possible responses:

Speak only English at home
Speak English very well *
Speak English well *
Do not speak English well *
Do not speak English at all *

(*those who do not speak only English at home)

NOTE: The variable ENGSPK5 in the datasets abbreviates these responses as: EngOnly, Very well, Well, Not well, and Not at all.

OTHER concepts

Race/Ethnicity (Topic two)   **Education** (Topic two)
Latino Groups (Topic two)   **Occupation** (Topic two)
Asian Groups (Topic two)   **Earnings** (Topic two)
State (Topic one)

A. The Immigrant Population

Over the last two decades, the race/ethnicity composition of the United States has become more diverse. In fact, the United States is now one of the most multi-ethnic societies in the world. Immigration, among other factors, plays a significant role in this growing diversity.

The foreign born population has increased absolutely and relatively since 1970. As a result of this influx, the immigrant population reached eight percent of the whole population by 1990. This trend in immigration is noteworthy not only for its rise in numbers, but also for the shift in its source regions. From the nineteenth to the first half of the twentieth century, the flow of immigrants came primarily from Canada and Europe. However, the proportion of these immigrants has decreased steadily over the last four decades. Since the 1970s, Asians and Latinos have been the largest growing immigrant groups in the United States.

Exercise 1 To set the stage for the issues you will explore in this chapter, look at the immigration status distribution of each race/ethnic group in 1990. What general patterns do you find? (*POPUSA9.DAT*)

■ Create a pie chart for each race/ethnic group. In each pie, make divisions for each immigration status.

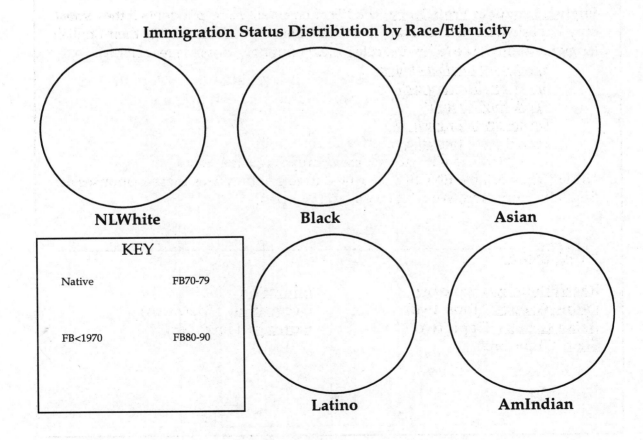

Immigration Status Distribution by Race/Ethnicity

NLWhite   Black   Asian

KEY

Native   FB70-79

FB<1970   FB80-90

Latino   AmIndian

Exercise 2 Now look at the immigration status of each specific Asian group. Which groups have the most native born members? Foreign born? Which groups have the earliest immigrants? Most recent? Why do you think this is so? (*ASNUSA9.DAT*)

■ Create a stacked bar chart with bars for each specific Asian group; stack by native born, immigrated before 1970, immigrated between 1970 and 1979, and immigrated between 1980 and 1990.

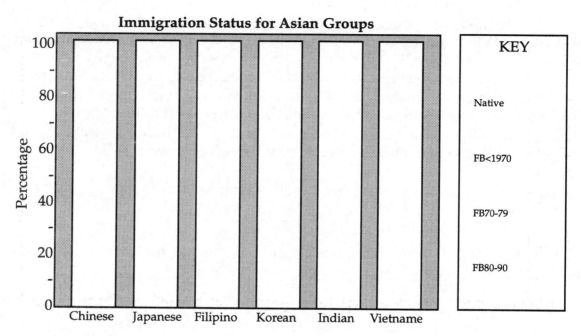

Exercise 3 Repeat the previous exercise for each specific Latino group. (*LATUSA9.DAT*)

■ Create a stacked bar chart with bars for each specific Latino group; stack by native born, immigrated before 1970, immigrated between 1970 and 1979, and immigrated between 1980 and 1990.

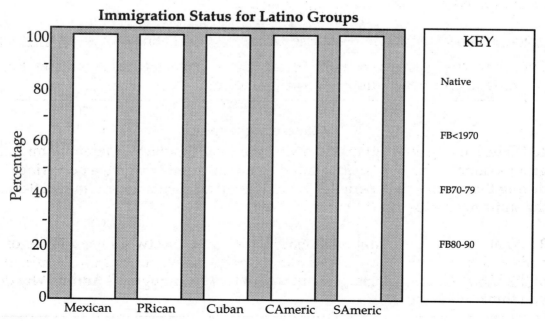

Exercise 4 Looking at people who have immigrated to the United States since 1980, examine gender differences for each race/ethnicity. What differences do you find between the race/ethnic groups? *(IMMUSA9.DAT)*

■ Create a bar chart with side by side bars for men and women; for each race/ethnic group, indicate the number of immigrants since 1980.

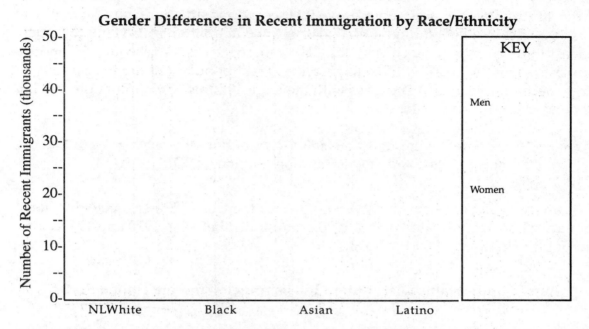

Gender Differences in Recent Immigration by Race/Ethnicity

Exercise 5 Now look at the differences in numbers between men and women in specific Asian groups who have immigrated to the U.S. since 1980. What differences in numbers do you find between males and females? Groups? *(ASNUSA9.DAT)*

■ On your own, create a bar chart with side by side bars for males and females; for each group, indicate the number of immigrants.

Exercise 6 Repeat the previous exercise for each specific Latino group. *(LATUSA9.DAT)*

■ On your own, create a bar chart with side by side bars for males and females; for each group, indicate the number of immigrants.

---

*D*iscussion **Questions**

1. While looking at the graphs showing the years in which different immigrant groups came to the U.S., keep in mind historical events that were occurring during the various time periods. Which historical events do you think affected the shifts in immigration?

2. What factors do you think influence the difference between the number of male and female immigrants? Conditions of the native countries? Conditions of the U.S.? Where is the largest number of female immigrants from? Why do you think this is the case?

B. Immigrant Geographic Location

Many immigrants settle near earlier immigrants from their country of origin. This established support network can provide stability and crucial contacts for immigrants arriving in a completely new world. Since many people immigrate to the United States in order to improve their economic situation, the availability of jobs also affects where immigrants choose to settle. Immigrants also tend to settle, at least initially, in ports of entry. Taking all of these factors into account, it is not surprising that the areas with the largest immigrant populations are in or near large metropolitan areas. The cities with the largest immigrant populations are Los Angeles, New York and Miami.

Exercise 7 Compare the immigration status distribution of Los Angeles to that of the United States in 1990. Are the distributions similar? *(POPUSA9.DAT, POPLA9.DAT)*

■ Create two pie charts, one for L.A. and one for the United States; make divisions for native born, immigrated before 1970, immigrated between 1970 and 1979, and immigrated between 1980 and 1990.

Immigration Status Distribution in Los Angeles and the United States

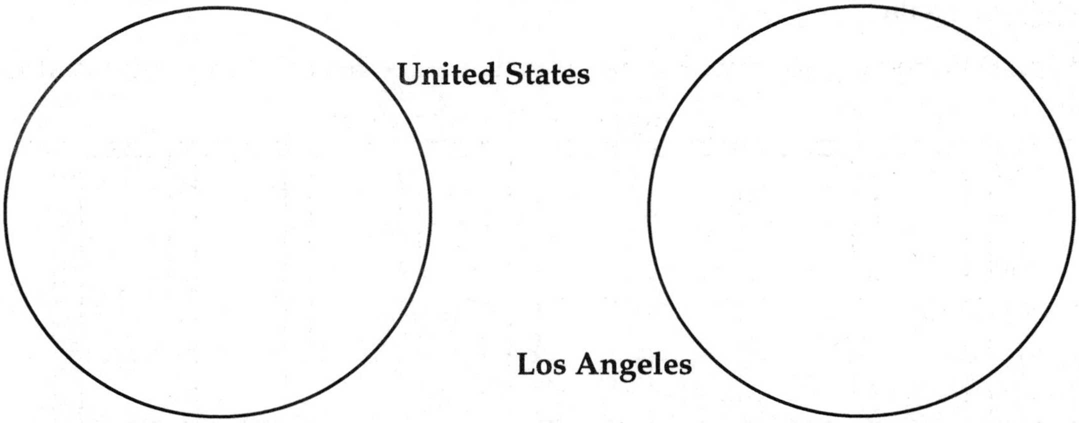

United States

Los Angeles

K	Native	FB<1970
E		
Y	FB70-79	FB80-90

Exercise 8 Now look at the immigration status distribution of Los Angeles for each race/ethnic group. Which race/ethnic groups have a greater number of recent immigrants? Earliest immigrants? *(POPLA9.DAT)*

■ Create a stacked bar chart for L.A. In the chart, make bars for each race/ethnic group; stack by immigration status.

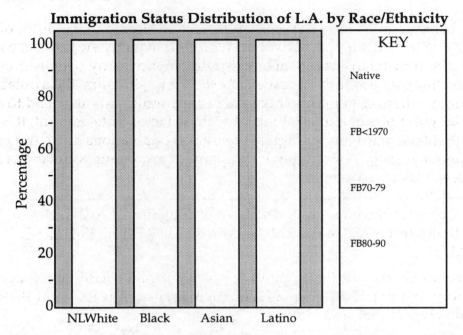

Immigration Status Distribution of L.A. by Race/Ethnicity

Exercise 9 Now look at the immigration status distribution in specific states in 1990. Which states have the largest percentages of recent immigrants? Earliest immigrants? (*IMMUSA9.DAT*)

■ Create a stacked bar chart with bars for each state; stack by immigration status.

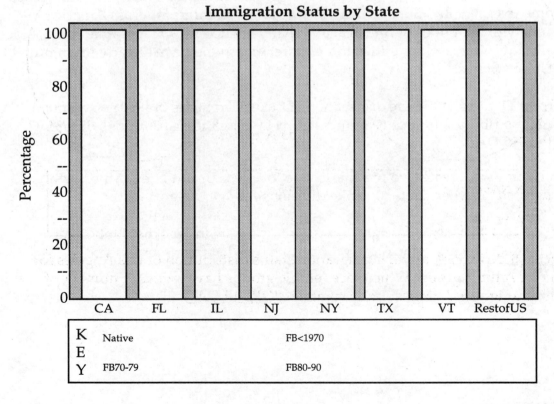

Immigration Status by State

Exercise 10 On your own, explore how the recent immigration has affected the race/ethnicity composition of certain states and the United States. (*IMMUSA9.DAT*)

$\mathcal{D}$iscussion Questions

1. Why is the immigrant population concentrated in particular areas? Also, explain why you think certain states have a small immigrant population.

2. Look at patterns of where immigrants settled prior to 1970 and after 1980. Does there seem to be a shift in where immigrants settle? What might account for this shift?

$C.$ Immigrant Age Structure

The age distribution of immigrants has a direct effect upon American society in areas such as fertility, health care, education, and employment. For example, many Asian immigrants arrive during their childbearing years and will have children after settling in the United States. As a result, the Asian population in the United States will inevitably grow.

An immigrant's age upon arrival also offers clues as to why he or she chose to come to the United States. Generally, younger immigrants cross American borders with the hopes of financial and material success. On the other hand, older immigrants often seek reunification with relatives who have already immigrated to the United States.

Exercise 11 Look at the age distribution of the United States in 1990 and compare it to the age distribution of recent immigrants. How are the distributions different? (*POPUSA9.DAT*)

■ On your own, create two pie charts, one for the U.S. and one for those who immigrated between 1980-90; make divisions for age categories.

Exercise 12 Focusing on foreign born Asians, compare the age distributions of immigrants in each specific Asian group in 1990. What trends are evident? (*ASNUSA9.DAT*)

■ Create a stacked bar chart with bars for immigrants of each Asian group; stack by ages 0-14, 15-34, 35-64, and 65+. (Hint: Combine ages.)

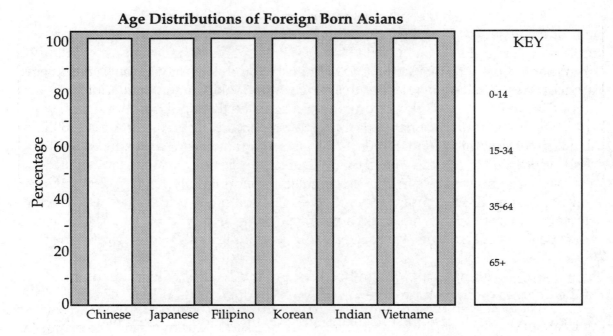

Age Distributions of Foreign Born Asians

KEY

0-14

15-34

35-64

65+

Chinese Japanese Filipino Korean Indian Vietnam

Exercise 13 Compare the 1990 age distributions of foreign born persons among specific Latino groups. What trends do you notice from the graphs? (*LATUSA9.DAT*)

■ On your own, create a stacked bar chart with bars for immigrants of each Latino group; stack by ages 0-14, 15-34, 35-64, and 65+.

---

*D*iscussion **Questions**

1. Is the age distribution for those who immigrated to the U.S. between 1980 and 1990 similar to the age distribution of the U.S.? Why do you think this is so? How do you think earlier immigrants' age distributions differed?

2. While looking at the Asian and Latino immigrant age distributions, consider why people arrive at the age they do. Do you think the reasons vary for immigrants of different ages? For example, think of political asylum, work opportunities, and family reunification.

𝒟. English Language Proficiency

An immigrant's command of English may affect his or her occupation, earnings and adjustment to living in the United States. While most acknowledge that proficiency in English is an important tool for immigrants, some Americans feel that immigrants should be required to speak English. Some states have proposed reducing services offered in languages other than English. By looking at English proficiency among immigrant populations, you can ascertain who these policies may affect.

English proficiency has been found to vary according to an immigrant's origin. To a large extent, this variation reflects which immigrant groups are exposed to English before they live in the United States. Duration of residence in the United States also plays a significant role in English language proficiency among immigrants. Generally speaking, English proficiency increases as duration of residence increases. However, an immigrant's age upon arrival, education level, and proximity to others who speak his or her native language all affect an immigrant's English proficiency.

Exercise 14 Looking at all foreign-born Asians and Latinos, show the distribution of the degrees of English proficiency. *(ENGLAT9.DAT, ENGASN9.DAT)*

■ Create two pie charts, one for Asians and one for Latinos. In each pie, make divisions for English proficiency categories.

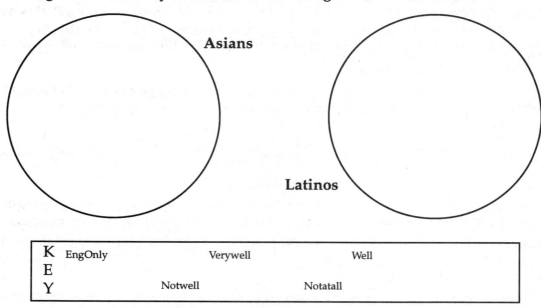

English Proficiency Distribution of Foreign Born Asians and Latinos

Exercise 15 Now look at the English language proficiency distribution in terms of specific Asian groups. Which group has the greatest percentage of those who speak English well? Which group has the greatest percentage of those who do not speak English well or not at all? What might account for these differences? *(ENGASN9.DAT)*

■ Create a stacked bar chart with bars for each Asian group; stack by language proficiency.

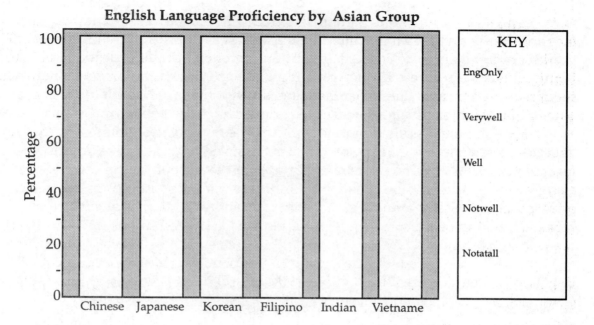

English Language Proficiency by Asian Group

Exercise 16 On your own, repeat the pevious exercise for specific Latino groups. (*ENGLAT9.DAT*)

Exercise 17 Now consider the duration of residence of all foreign-born individuals of Asian and Latino groups while examining the distribution of English proficiency. How does English proficiency change as groups are in the U.S. longer periods of time? (*ENGASN9.DAT, ENGLAT9.DAT*)

■ On your own, create two stacked bar charts, one for Asians and one for Latinos. In each chart, create bars for each immigrant status category; stack by language proficiency.

---

*D*iscussion Questions

1. Recall the exercise in which you looked at language proficiency of immigrants from different Asian or Latino groups. Which group has the highest percentage of proficient English speakers? Why do you think this is the case? Cultural factors? Educational factors?

2. Now turn your attention to the exercise in which you looked at the language proficiency across different immigrant statuses. What factors do you think contribute to such a distribution? Increased education? Attempts at assimilation? Was it consistent that the longer an immigrant is in the country, the more proficient that person's English became? If not, provide some possible explanations for why this might be so.

ℰ. Education and Assimilation

Immigrants' education levels show more extremes than the general, native-born population. On the one hand, compared with the native-born, immigrants are more likely to have Ph.D's and higher levels of education. On the other hand, they are more likely to be concentrated at the lower end of the education spectrum. Many of these differences have to do with the selectivity patterns of immigrants at their country of origin. Some of this has to do with the immigration laws in the United States which give preferences to certain categories like family reunification, or those with needed skills for jobs that cannot be filled by existing residents. In general, there are more immigrants who are less well-educated in comparison with those at the other extreme. The investigations in this section will allow you to examine these differences, overall and for different immigrant groups. You will also determine whether the most recent immigrants—those who arrived since 1980—are less well-educated than immigrants who arrived earlier.

Exercise 18 Look at the educational attainment distribution of the entire U.S. population in 1990 for individuals ages 25-34. (*EDUCIMM9.DAT*)

■ Create a pie chart with divisions for education levels.

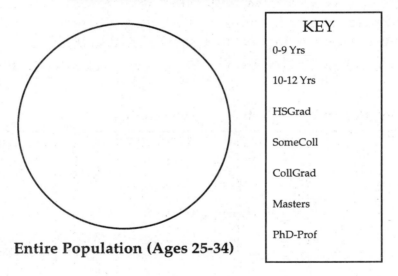

Educational Distribution of the U.S.

KEY

0-9 Yrs

10-12 Yrs

HSGrad

SomeColl

CollGrad

Masters

PhD-Prof

Entire Population (Ages 25-34)

Exercise 19 Now look at the educational attainment differences between foreign born individuals, ages 25-34, in specific Asian groups. Which groups have the largest percentage of people with at least a college degree? The lowest? (Hint: Combine all foreign born categories.) (*EDUASN9A.DAT*)

■ Create a stacked bar chart with bars for each group; stack by education level.

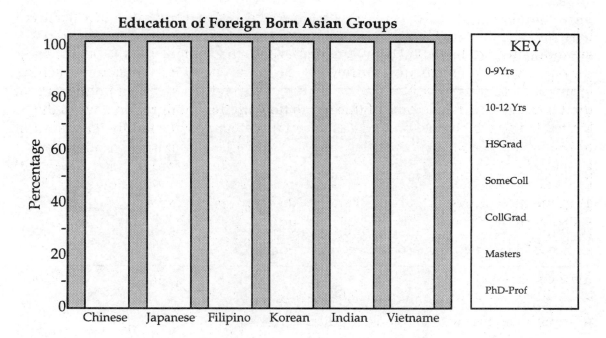

Education of Foreign Born Asian Groups

Exercise 20 Now take a look at the educational attainment differences between foreign born individuals, ages 25-34, in specific Latino groups. Which groups have the highest percentage of people with at least a college degree? The lowest? Compare your findings to the previous exercise. (*EDULAT9A.DAT*)

■ Create a stacked bar chart with bars for each group; stack by education level.

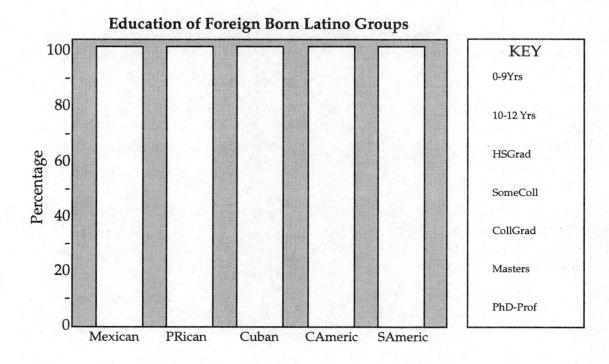

Education of Foreign Born Latino Groups

Exercise 21 Is the educational attainment of immigrants influenced by their duration of residence in the United States? On your own, compare the education distribution of recent immigrants to those who came before 1970. Look at this relationship in a few race/ethnic groups of your choice. Does the relationship between education and immigrant status vary between race/ethnic groups? (*EDUCIMM9.DAT*)

---

*D*iscussion Questions

1. Consider the educational differences you found between specific Asian groups and specific Latino groups. What do you think could explain these differences? Cultural factors? Economic factors?

2. Consider the growing need in the U.S. for skilled and unskilled labor. Which needs are met by recent immigration patterns? Do you think labor force needs have an influence on immigration policy?

F. Occupation and Assimilation

Differences between immigrants' occupations often reflect disparities in educational attainment and English language proficiency, two factors you have already considered. A large number of immigrants who have professional and managerial jobs are from Asia and Europe. Whereas Latinos comprise a significant proportion of immigrants who have labor and service positions.

In addition to looking at source regions, it is important to look at gender differences. Male immigrants are heavily concentrated in service and agricultural jobs. Furthermore, they have the lowest representation in managerial and sales positions. Female immigrants tend to concentrate in service and blue collar employment.

Exercise 22 Focusing on men ages 25-34, examine the occupational distribution of the entire U.S. population in 1990. Compare it to the distribution of those who immigrated to the U.S between 1980 and 1990. How are the distributions different? (*OCIM9-25.DAT*)

■ Create two pie charts, one for the U.S. and one for recent immigrants. Make divisions for occupation categories.

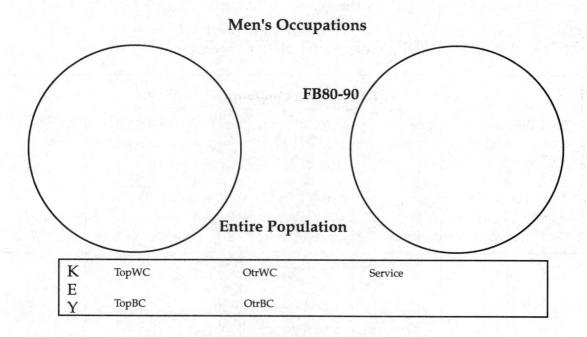

Men's Occupations

FB80-90

Entire Population

K E Y	TopWC	OtrWC	Service
	TopBC	OtrBC	

Exercise 23 Still focusing on 25-34 year old males, consider the ways in which occupational distribution changes as immigrants have been in the U.S. for longer periods of time. What trends are evident? *(OCIM9-25.DAT)*

■ Create a stacked bar chart with bars for each immigration status; stack by occupation.

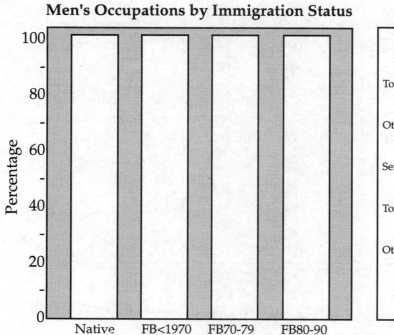

Men's Occupations by Immigration Status

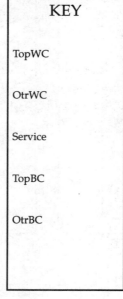

KEY

TopWC

OtrWC

Service

TopBC

OtrBC

Exercise 24 Look at the occupational distribution of 25-34 year old male and female Asians. Compare the occupational distribution of recent immigrants to that of earlier immigrants. What are the gender differences? *(OCIM9-25.DAT)*

■ Create two stacked bar charts, one for men and one for women. In each chart, draw bars for each immigration status; stack by occupation.

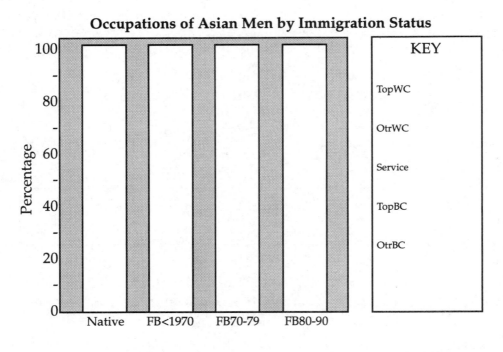

Occupations of Asian Men by Immigration Status

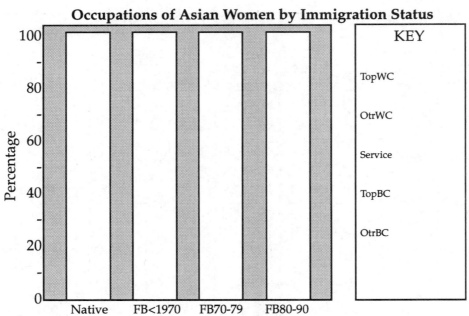

Occupations of Asian Women by Immigration Status

Exercise 25 On your own, repeat the previous exercise for Latinos. How do Latinos compare to Asians in terms of occupational distribution? *(OCIM9-25.DAT)*

Exercise 26 Now examine the occupational distributions of Asian and Latino immigrants for each education level. You may focus on either females OR males, ages 25-34. As education levels increase, what happens to the occupational distribution? Does the relationship between education and occupation vary between Asians and Latinos? *(OCIM9-25.DAT)*

■ Create two stacked bar charts, one for Asians and one for Latinos. In each chart, draw bars for each education level; stack by occupation. Specify whether you looked at males or females.

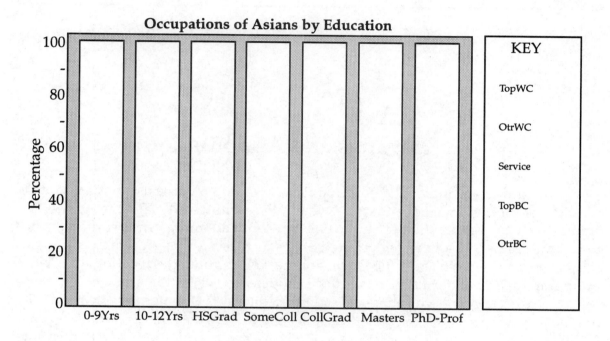

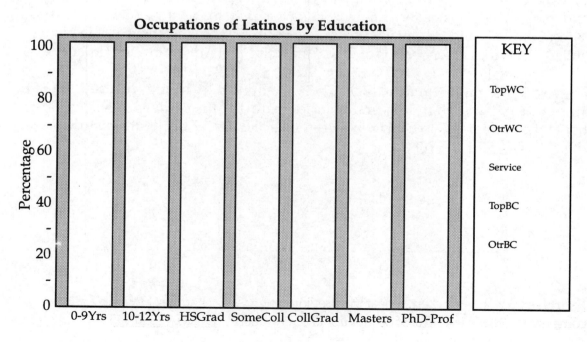

𝒢. Earnings and Assimilation

In addition to indicating economic status, earnings serve as an important measure of immigrants' adjustment. Overall, the earnings of male immigrants is significantly lower than that of native born men, but an immigrant's earnings tend to increase as his duration in the United States increases. This growth in earnings reflects greater experience in the U.S. labor market, continued education following immigration, and better English language proficiency.

Like occupation, education and English proficiency, earnings vary by region of origin. Immigrants from Europe and Canada tend to have the highest earnings, followed closely by the Asian immigrants. The earnings of Latin American immigrants tend to be lower. This pattern might be linked, in part, to the duration in the United States. European and Canadian immigrants, generally speaking, have resided in the United States for longer periods of time than their Latin American counterparts.

Exercise 27 Focusing on males ages 25-34, examine the earnings distribution of the entire U.S. population in 1990 and then compare it to the earnings distribution of those who immigrated to the U.S. between 1980 and 1990. How do the distributions differ? (*WKIM9-25.DAT*)

■ Create two pie charts, one for the U.S. and one for recent immigrants. In each pie, make divisions for earnings categories.

Earnings Distributions

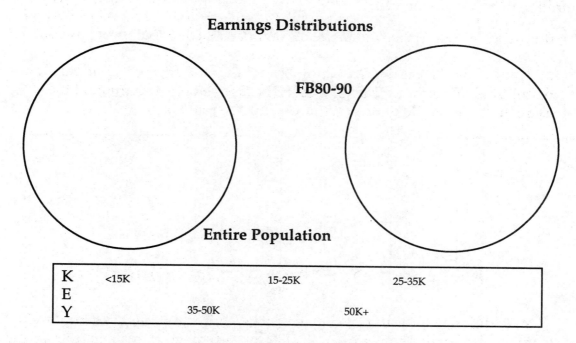

FB80-90

Entire Population

K E Y	<15K	15-25K	25-35K
	35-50K	50K+	

Exercise 28 Now look specifically at male college graduates, ages 25-34, in top white collar occupations. Compare the earnings of recent immigrants (1980-1990) to the native born population. Do the two groups show similar distributions? Why do you think this is so? (*WKIM9-25.DAT*)

■ Create two pie charts, one for recent immigrants and one for native born Americans. In each pie, make divisions for earnings categories.

Earnings Distributions of College Graduates, Ages 25-34

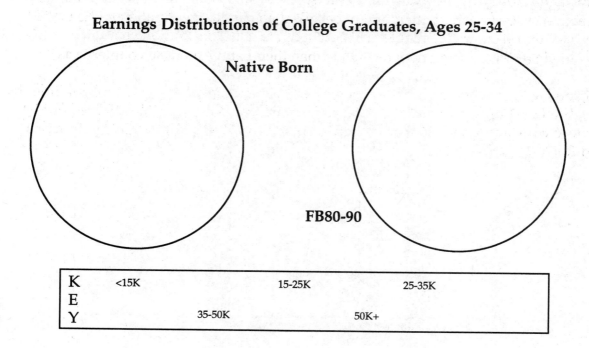

Native Born

FB80-90

K E Y	<15K	15-25K	25-35K
	35-50K	50K+	

Exercise 29 Focusing on Latino men, ages 35-44, show the earnings distribution while taking immigration status into account. Do earnings consistently increase as duration of residence increases? (_WKLT9-35.DAT_)

■ Create a stacked bar chart with bars for each immigration status category; stack by earnings.

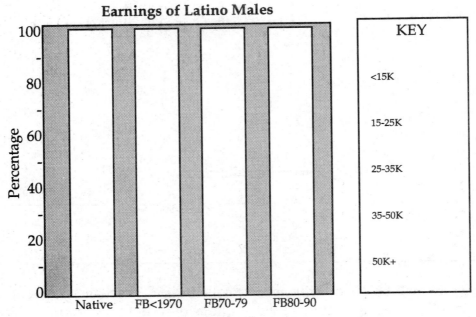

Exercise 30 Determine the earnings distribution of Latino males, ages 35-44, in top white collar positions, in terms of immigration status. What trends do you notice among those who earn more than 25K a year? (_WKLT9-35.DAT_)

■ Create a stacked bar chart for Latino top white collar workers with bars for each immigration status category; stack by earnings.

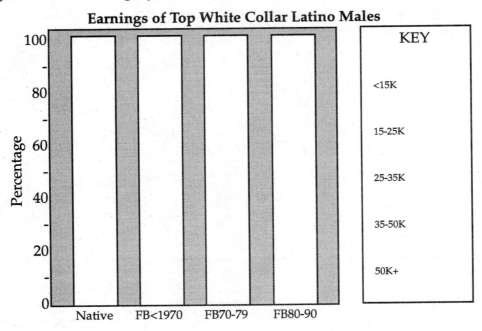

Exercise 31 Now look at Asian men, ages 35-44. Show the earnings distribution while taking immigration status into account. Do earnings consistently increase as duration of residence increases? Compare your findings to exercise 29. (*WKAS9-35.DAT*)

■ Create a stacked bar chart with bars for each immigration status category; stack by earnings.

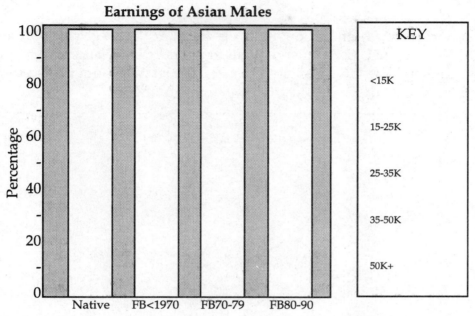

Exercise 32 Determine the earnings distribution of Asian males, ages 35-44, in top white collar positions, in terms of immigration status. What trends do you notice among those who earn more than 25K per year? (*WKAS9-35.DAT*)

■ Create a stacked bar chart for Asian top white collar workers with bars for immigration status category; stack by earnings.

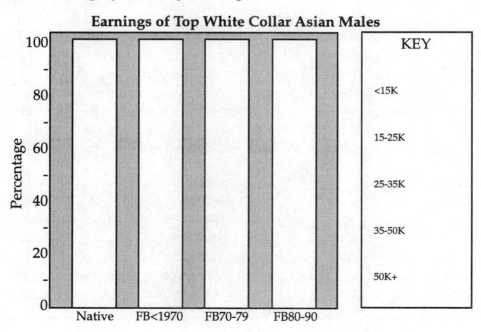

$\mathcal{D}$iscussion Questions

1. How do you think occupation, immigration status, and education affect the earnings of immigrants? Do you think these factors, with the exception of immigration status, affect native born residents in the same fashion? Why or why not?

2. Imagine you have a friend from one of the Asian or Latino groups studied in this text. They see the U.S. as a "land of golden opportunity" and plan to move here to make their "fortune". Using what you have learned from this section as well as others, what advice would you give to this friend?

THINK tank

1. Since most cultures have unique conceptions of gender and occupational roles, examine whether occupation, education, and earnings patterns are different for female and male immigrants. What similarities and differences do you see between native born and foreign born females? Does the labor market appear to be biased against immigrants in general? Specifically against females? Use the data to explain and defend your answer.

2. Some people think the U.S. has too many immigrants who get too many government *hand-outs*. Other people argue that immigrants *pay for themselves*, and make a positive contribution to our economy. Do you think immigration is good for the U.S.? Should current policies be changed to allow more or fewer immigrants? Why or why not? What are the potential social, economic, and political implications of current immigration conditions? Use the data to make inferences and support your position.

LABOR FORCE *topic four*

How many times has someone asked you, "What do you do?" or "What are you going to do after you graduate?"? By *do*, the questioner usually means *work*. It is not surprising that this question surfaces repeatedly because many Americans consider their work an integral part of who they are.

In light of the role of work in our lives as both an economic and personal achievement indicator, it is important to take a look at labor force participation trends. A person is considered to be in the labor force if he or she is currently working or looking for work. Persons not in the labor force typically include students, homemakers, retirees and disabled people.

Of those in the labor force, the percentage of unemployed persons varies greatly between gender, race/ethnic, and age groups. Yet, all of these groups' unemployment trends are strongly affected by periods of economic growth and recessions. In the early 1950s, unemployment rates were quite low, but rose during a recession in the late 50s. The "guns and butter" period of the late 60s saw low unemployment levels which sharply increased during the 1973-1975 recession period. Similar fluctuations have occurred throughout the 1980s and early 90s as workers scramble to

acquire new skills to meet the needs generated by economic restructuring.

Although all groups have been affected by economic changes, some groups have been more affected than others. Usually, blacks suffer a sharper increase in unemployment than nonblacks during periods of economic decline. However, when the economy picks up again, unemployment rates drop faster among blacks than nonblacks. Economic changes also have a varying impact upon different age groups. Due to seniority status, older people can often avoid layoff during recession. On the other hand, since employers are less likely to hire someone who they think is going to retire soon, economic upswings have little impact upon older people's unemployment rates. A shift in labor market needs can cause different rates of unemployment between men and women.

While unemployment trends closely follow the peaks and valleys of the economy, trends in labor force participation are not as strongly aggected by eco-

KEY concepts

Labor Force Status The civilian labor force includes persons ages 16 and over who either have a job (employed) or are able and looking for work (unemployed). Therefore, the civilian population ages 16 and over can be classified into the following:

>*In Labor Force—Employed* persons with a full-time or part-time job.
>*In Labor Force—Unemployed* persons who are able to work and who are looking for work or laid off from a job
>*Not in the Labor Force* Persons without a job and not available for work (e.g. retirees, homemakers, full-time students, etc.)

NOTE: These categories are abbreviated as Empd, Unempd, and NILF for variable EMP3 in datasets.

Percent in Labor Force (Labor Force Participation Rate)
Calculated from the above categories as:
$$\frac{\text{Employed + Unemployed}}{\text{Employed + Unemployed + Not in Labor Force}} \times 100$$

Percent Unemployed (Unemployment Rate)
Calculated from above categories as:
$$\frac{\text{Unemployed}}{\text{Employed + Unemployed}} \times 100$$

Full-time - Part-time Workers Employed workers are considered to be full-time workers if they usually work more than 35 hours per week. Part-time workers work less than 35 hours but can be classed more specifically by the number of weekly hours they usually work.

NOTE: The WKHRS4 variable in the datasets classes workers according to the following: *Full35* (full-time workers), *20-34*, *10-19*, and *Under 10* (different usual weekly hours for part-time workers). Also the EMP4 variable distinguishes between full-time and part-time workers with the categories *EmpFull* and *EmpPart*.

Year-round - Full-time Workers Employed workers who work more than 35 hours per week for 50 or more weeks per year.

OTHER concepts

Race/Ethnicity (Topic two)   **Marital Status** (Topic five)
Education (Topic two)   **Children Ever Born** (Topic five)

changes, but are shaped by societal forces. Women's labor force participation has risen markedly since the 1960s due to choice and necessity. Elderly men show a greater likelihood to withdraw from the labor force during their late 50s and 60s. By completing the following exercises, you will learn how labor force participation and employment vary across different groups in society.

$\mathcal{A}.$ Labor Force Trends

Labor force participation rates have changed in important ways since 1950. These changes have been especially dramatic for women and older men. Despite the Civil Rights movement and other societal changes, race/ethnicity gaps in labor force participation are increasing among men. On the other hand, black women have had a unique position in the labor force for several decades.

In the following exercises, you will look at labor force participation and unemployment in terms of gender, race/ethnicity, and age. Since women exit and reenter the labor force more often than men, and these fluctuations would skew overall trends, it is necessary to look at men and women's labor force participation separately.

Exercise 1 How has the percentage of men and women in the labor force changed over time? Using data from 1950 to 1990, look at the percentage of men and women, ages 16 and older, in the labor force. Describe the trends. (*EMP5090.DAT*)

■ Create a line graph with two lines, one for men and one for women, ages 16 and older. For each year, indicate the percentage of those in the labor force. (Hint: Combine Empd with Unempd as discussed in "Key Concepts".)

Men and Women in the Labor Force, 1950 to 1990

Exercise 2 How has the percentage of unemployed men and women changed over time? Using data from 1950 to 1990, look at the percentage of men and women, ages 16 and older, unemployed. What trends do you notice? (*EMP5090.DAT*)

■ Create a line graph with two lines, one for men and one for women, ages 16 and older. For each year, indicate the percentage of those unemployed. (Hint: Omit NILF.)

Exercise 3 Now look at the relationship between race and the percentage of men and women in the labor force. Using data from 1950 to 1990, look at the percentage of black and nonblack men and women, ages 16 and older, in the labor force. How do these trends differ from those in the previous exercises? (*EMP5090.DAT*)

■ Create a line graph with four lines, one for black men, one for nonblack men, one for black women, and one for nonblack women. For each year, indicate the percentage of those in the labor force.

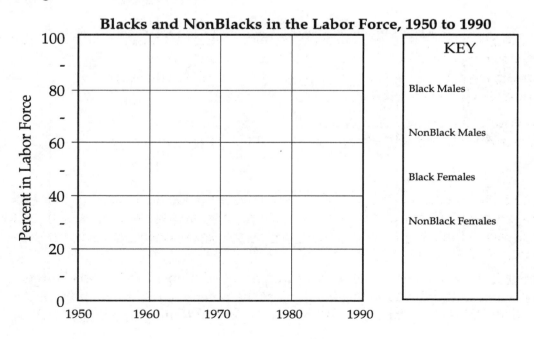

Exercise 4 What is the relationship between race and unemployment of men and women? Using data from 1950 to 1990, look at the percentage of unemployed black and nonblack men and women, ages 16 and older. How do these trends differ from those in the previous exercises? (*EMP5090.DAT*)

■ Create a line graph with four lines, one for black men, one for nonblack men, one for black women, and one for nonblack women. For each year, indicate the percentage of those unemployed.

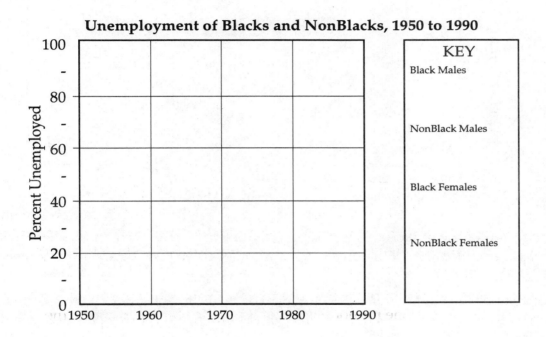

Exercise 5 Focusing on 1990, examine the relationship between gender, race/ethnicity and the percentage of people in the labor force. Looking at men and women, ages 16 and older, in each race/ethnic group, what differences do you see between genders and race/ethnicities? (*EMPED9.DAT*)

■ On your own, create a bar chart with side by side bars for men and women. For each race/ethnic group, indicate the percentage of men and women in the labor force.

Exercise 6 Focusing on 1990, examine the relationship between race/ethnicity and unemployment of men and women. Looking at men and women, ages 16 and older, in each race/ethnic group, what differences do you see between genders and races/ethnicities? (*EMPED9.DAT*)

■ Create a bar chart with side by side bars for men and women. For each race/ethnic group, indicate the percentage of men and women unemployed.

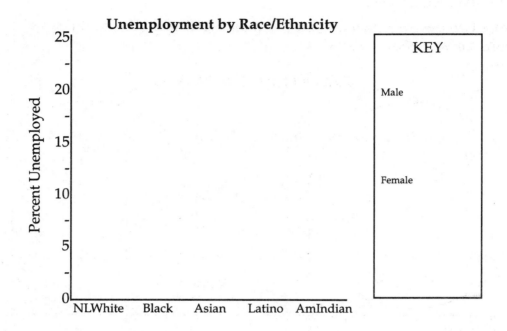

Exercise 7 Looking at 1950 and 1990 separately, determine the percentage of women in each age group in the labor force. What are the differences between these years? (*EMP5090.DAT*)

■ Create two bar charts, one for 1950 and one for 1990. In each chart, draw a set of age group bars for women. In each chart, for each age group, indicate the percentage in the labor force.

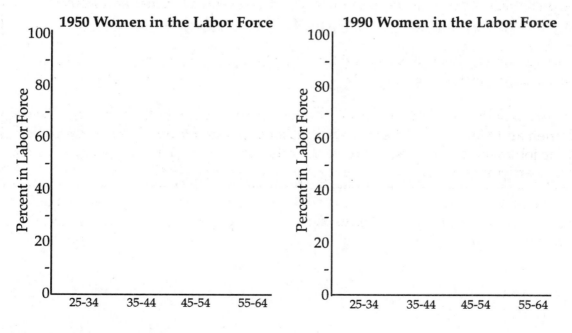

Exercise 8 Does gender influence an individual's work hours? Focusing on men and women, ages 16 and older, in the labor force in 1990, look at their full-time/part-time status. What might account for the difference between men and women's work hours? *(FPRTED9.DAT)*

■ Create two pie charts, one for men and one for women. In each pie, make divisions for work hour status.

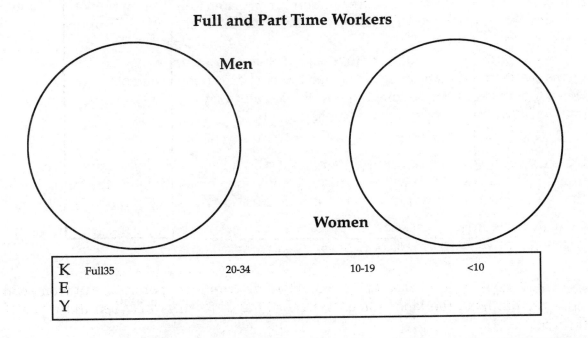

Full and Part Time Workers

K	Full35	20-34	10-19	<10
E				
Y				

---

*D*iscussion **Questions**

1. What historical events may have affected unemployment trends and changes in the percentage of people in the labor force? Did these events affect men and women differently?

2. In addition to race/ethnicity and gender, what other factors may affect an individual's employment status?

3. Compare the labor force participation trends of black women to those of white women and black men. Offer possible reasons for black women's unique history in the job market. What factors have contributed to their changing levels of labor force participation, both overall and relative to other groups?

$\mathscr{B}$. Men's Labor Force Participation

While women have experienced a noticeable increase in labor force participation, there has been a decrease in men's labor force participation. Overall, this decrease has not been especially dramatic, but some groups have been affected more than others. Changes in social security, pension, and early retirement have led to a reduction in older men's labor force participation. At the same time, young black men have experienced both a decrease in labor force participation and an increase in unemployment.

As you work through the following exercises, consider why some groups of men have more dramatic decreases in labor force participation than other groups. When looking at unemployment trends, think about why certain economic changes may have affected some groups more than others.

Exercise 9 Look at the percentage of black males, ages 16-24, in the labor force from 1950 to 1990. How has the percentage of young black males in the labor force changed over time? What might account for these changes? (*EMP5090.DAT*)

■ Create a line graph indicating the percentage of black males, ages 16-24, in the labor force for each year.

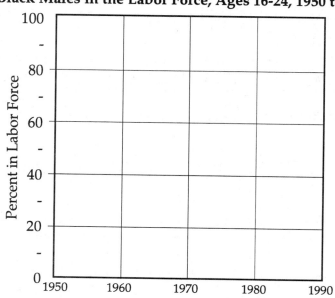

Black Males in the Labor Force, Ages 16-24, 1950 to 1990

Exercise 10 Look at the percentage of black males, ages 16-24, unemployed from 1950 to 1990. How has the percentage of unemployed young black males changed over time? What might account for these changes? (*EMP5090.DAT*)

■ Create a line graph indicating the percentage of black males, ages 16-24, unemployed for each year.

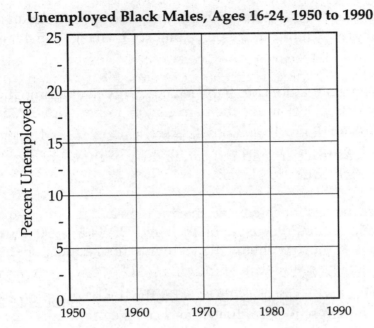

Unemployed Black Males, Ages 16-24, 1950 to 1990

Exercise 11 Focusing on 1990, look at men ages 16-24, 35-44, and 55-64 in each race/ethnic group. What percentage of each group is in the labor force? Between which age groups and races/ethnicities do you see significant differences? (*EMPED9.DAT*)

■ Create a bar chart with side by side bars for the three age groups. For each race/ethnic group, indicate the percentage of each age group in the labor force.

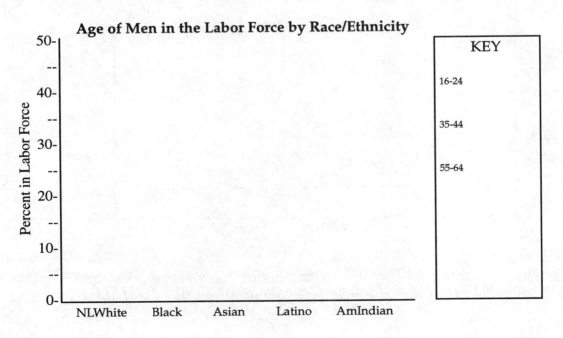

Age of Men in the Labor Force by Race/Ethnicity

KEY

16-24

35-44

55-64

Exercise 12 Now focus on 35-44 year old male Asians and Latinos in the labor force in 1990. Looking at each specific Asian and Latino group, note any significant differences. *(EMPASN9.DAT, EMPLAT9.DAT)*

■ On your own, create a bar chart with bars for each specific Asian and Latino group. For each group, indicate the percentage of men, ages 35-44, in the labor force in 1990.

Exercise 13 Is there a relationship between educational attainment and unemployment? Focusing on 1990, examine the percentage of unemployed 35-44 year old men for each education level. *(EMPED9.DAT)*

■ On your own, create a bar chart indicating the percentage of men, ages 35-44, unemployed in each education level.

Exercise 14 Does the relationship between education and unemployment vary between race/ethnic groups? Focusing on men ages 35-44 in 1990, look at unemployment percentages of those with less than a high school education, high school graduates, and college graduates. *(EMPED9.DAT)*

■ Create a bar chart with side by side bars for the three education categories: less than high school, high school graduates, and college graduates. For each race/ethnic group, indicate the percentage of 35-44 year old men unemployed in each education category.

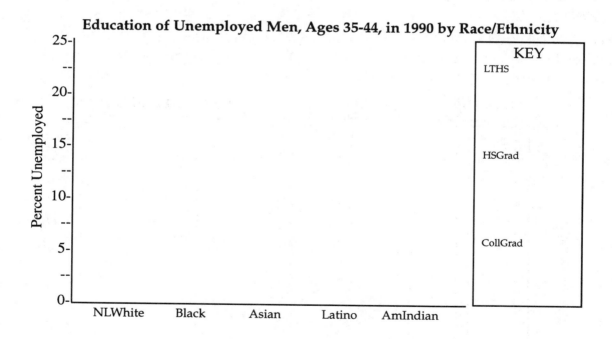

Exercise 15 Using 1990 data, look at the full-time/part-time work status of males in each age group. How does the work hour status distribution vary? *(FPRTED9.DAT)*

■ Create a stacked bar chart with bars for each age group; stack by work hour status.

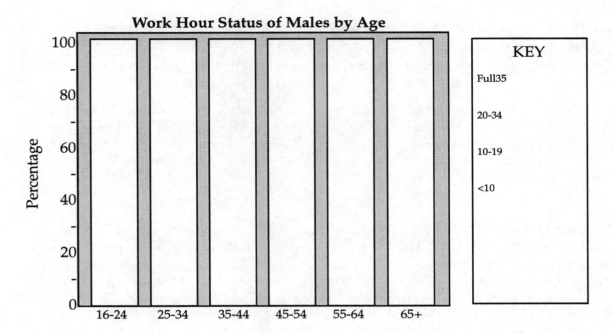

Work Hour Status of Males by Age

Exercise 16 Is there a relationship between race/ethnicity and full-time/part-time work status? Using 1990 data, look at the work hour status of males ages 35-44 in each race/ethnic group. How does the hourly status distribution vary? (*FPRTED9.DAT*)

■ Create a stacked bar chart with bars for each race/ethnic group; stack by work hour status.

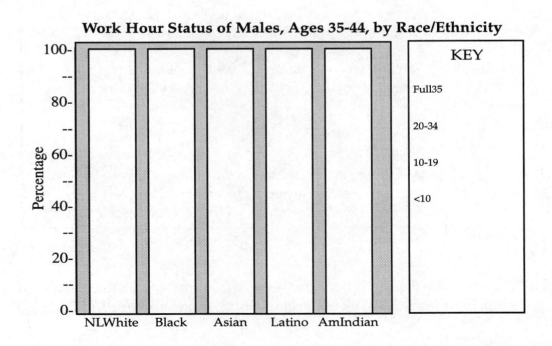

Work Hour Status of Males, Ages 35-44, by Race/Ethnicity

Discussion Questions

1. Think about unemployment among young black males. What are some possible reasons for this group's past and present position in the labor market? Explain why some economic factors may have had more of an impact on this group's job opportunities than on other race/ethnic groups.

2. Explore the differences between the labor force participation of Asian and Latino groups. Discuss these groups' positions in the labor market, both overall and relative to other race/ethnic groups. Offer possible reasons for the differences you found.

C. Women's Labor Force Participation

Over the last few decades, the participation of women in the labor force has increased dramatically. This surge is especially significant among middle class women, for working class women have always participated in the labor force.

Rising economic demands on families, coupled with a greater number of women who desire more independence, have led to an increase in the number of women who work outside their homes. Also, the impact of the women's movement has provided the support many women need in order to enter the labor force.

Although women are joining the labor force in greater numbers, they also exit and reenter the labor force more than men do. This pattern may be partially attributed to childbearing and family responsibilities. Lower earnings and fewer opportunities for advancement also play a role in a woman's likelihood to exit the labor force. At the same time, the relationship between childbearing and labor force participation has changed. A majority of women now work outside the home during their childbearing years.

In the following exercises, you will examine overall gender differences in employment status and labor force participation, as well as the gender differences among race/ethnic groups.

Exercise 17 How does the percentage of women in the labor force vary by age for each race/ethnic group? Focusing on 1990, look at the percentage of women, ages 16-24 and 25-34, in the labor force for each race/ethnic group. What are the significant differences? (*EMPLOY9.DAT*)

■ Create a bar chart with side by side bars for the two age groups. For each race/ethnic group, indicate the percentage of each age group in the labor force.

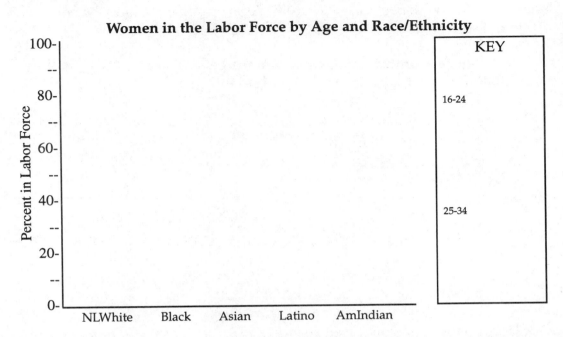

Women in the Labor Force by Age and Race/Ethnicity

Exercise 18 On your own, explore the relationship between unemployment, race/ethnicity and age among women in 1990. Repeat the previous exercise, but indicate the percentage unemployed rather than in the labor force. (*EMPLOY9.DAT*)

Exercise 19 Focusing on women ages 25-34 in 1990, look at the percentage in the labor force for each specific Latina group. What differences do you note? (*EMPLAT9.DAT*)

■ Create a bar chart with bars for each specific Latina group indicating the percentage of 25-34 year old women in the labor force.

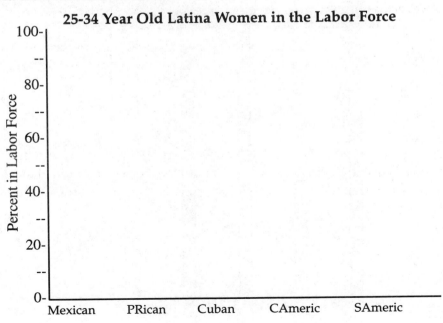

25-34 Year Old Latina Women in the Labor Force

Exercise 20 Focusing on women ages 25-34 in 1990, look at the percentage in the labor force for each specific Asian group. What differences do you note? (_EMPASN9.DAT_)

■ Create a bar chart with bars for each specific Asian group indicating the percentage of 25-34 year old women in the labor force.

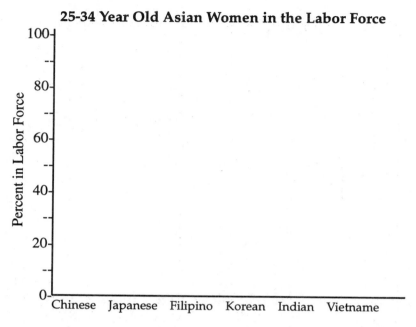

25-34 Year Old Asian Women in the Labor Force

Exercise 21 Using 1990 data, look at the hourly status of women in each age group. How does their full-time/part-time work status vary? What might account for these variations? (_FPRTED9.DAT_)

■ Create a stacked bar chart with bars for each age group; stack by work hour status.

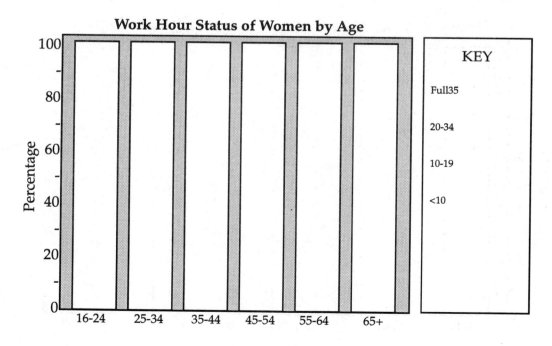

Work Hour Status of Women by Age

Exercise 22 Is there a relationship between race/ethnicity and full-time/part-time work status? Using 1990 data, look at the work hour status of females ages 35-44 in each race/ethnic group. What might account for the differences? *(FPRTED9.DAT)*

■ Create a stacked bar chart with bars for each race/ethnic group; stack by hourly status.

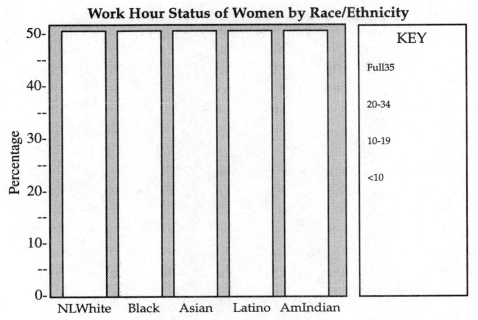

Work Hour Status of Women by Race/Ethnicity

Exercise 23 Is there a relationship between a woman's educational attainment and how many hours she works? Focusing on 1990, examine the work hour status of 25-34 year old women for each education level. What factors might affect how many hours a woman works? *(FPRTED9.DAT)*

■ Create a stacked bar chart with bars for each education level; stack by work hour status.

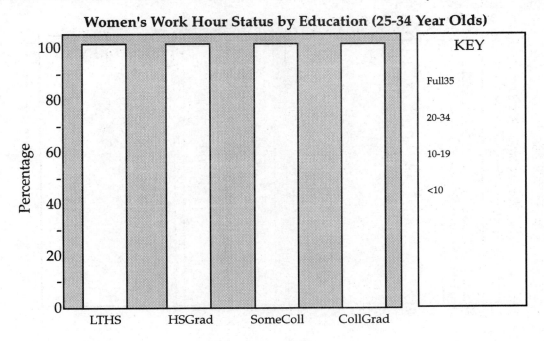

Women's Work Hour Status by Education (25-34 Year Olds)

Exercise 24 Is there a relationship between a woman's educational attainment and whether she is in the labor force? Focusing on 1990, examine the percentage of 25-34 year old women in the labor force for each education level. (*EMPED9.DAT*)

■ Create a bar chart with bars for each education level indicating the percentage of women, ages 25-34, in the labor force.

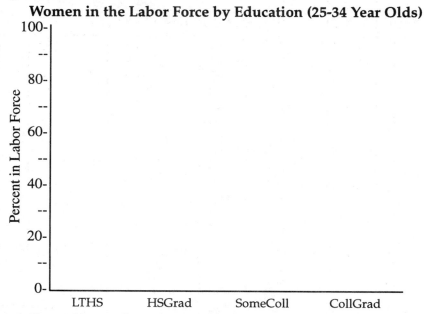

Women in the Labor Force by Education (25-34 Year Olds)

Exercise 25 Considering the relationship between a woman's educational attainment and whether she is in the labor force, look at the relationship between education and childbearing. Focus on women ages 25-34 in 1990. Compare the patterns to your findings in the previous exercise. (*BORN9.DAT*)

■ Create a stacked bar chart with bars for each education level; stack by the number of children ever born.

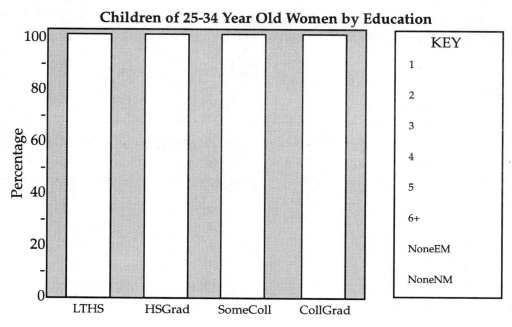

Children of 25-34 Year Old Women by Education

Exercise 26 Now look at the relationship between education and marital status. Focusing on women ages 25-34 in 1990, look at the marital status distribution for each education level. Compare the patterns to your findings in the previous exercises. (*MARED9.DAT*)

■ Create a stacked bar chart with bars for each education level; stack by marital status.

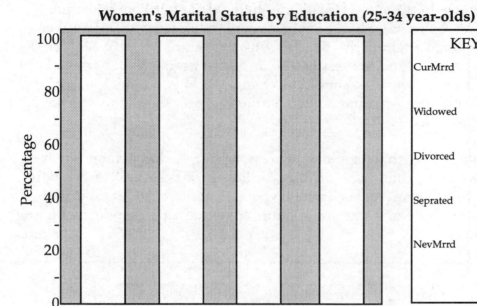

Women's Marital Status by Education (25-34 year-olds)

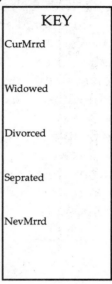

KEY

CurMrrd

Widowed

Divorced

Seprated

NevMrrd

$\mathcal{D}$iscussion **Questions**

1. Describe the overall trend in the percentage of women working since 1950. How does this trend compare to the employment patterns of men since 1950? How has the gap between the percentage of women and men in the labor force changed over the past several decades? What factors have influenced these changes?

2. Provide possible reasons for the labor force participation patterns of Latina, black, Asian, and American Indian women.

3. Considering your findings for 1990, predict the gender and race/ethnicity distribution of the labor force in 2010.

THINK tank

1. Many Americans believe that a good education leads to a stable job, and more education leads to higher paying jobs. While this seems to be the general trend, does more education always 'payoff' in the labor market? Do some individuals with fewer years of schooling earn more than those with higher levels of educational attainment? Explain why this might happen. Do you think schools, colleges, and universities do an adequate job of preparing people for the labor force? If not, what should be changed? Do you think the public education system prepares some types of students better than others? If so, who benefits the most? Why?

2. Some people believe that people who are poor simply don't want to work. Describe the working poor. Who are they? What are some other reasons why people who work or want to work may be poor or in poverty? Should individuals who work the same amount of hours enjoy the same standard of living? Why or why not?

In the 1950s, television shows like "Ozzie and Harriet" and "Leave It To Beaver" portrayed the typical family as consisting of two parents, married and living together, with at least two children. These TV shows were accurate representations of married life in this period. Marriage rates were already high and rising, divorce rates were low and the birth rate was skyrocketing.

While June and Ward Cleaver embodied the behaviors and values of post World War II America, many people mistakenly think this period was a mere continuation of earlier marriage and birth patterns. But this is not quite the case. The end of World War II precipitated a dramatic increase in births. Known as the Baby Boom, this trend continued into the early 1960s. During this period, five out of six women in peak childbearing years gave birth to at least two children. Americans were also marrying early and staying married.

As the baby boomers matured, they did not follow their parents' marriage and childbearing patterns. More and more people delayed marriage until their late twenties or early thirties. Couples delayed having children and had fewer children. Divorces increased as well.

As divorces became common and marriages were delayed, the number of people "cohabiting" — living together without being married — increased. Cohabitation is not only common among young people, but also an alternative for divorced people who are not ready to get remarried.

Americans' marriage choices have also changed. There has been an increase in the number of Americans marrying people of different race/ethnic backgrounds. As more people delay marriage or remarry later in their lives, more people tend to marry partners with similar education levels.

There no longer seems to be a "typical" marital life-style that can be portrayed by a single TV show. In the following exercises, you will look at marital status trends, marriage choices, divorce, cohabitation and childbearing decisions. Your findings will give you a clearer picture of how marriage in America is changing.

KEY *concepts*

Marital Status Classified according to the following:
    Currently married currently married and not separated
    Widowed widows and widowers who have not remarried
    Divorced legally divorced persons who have not remarried
    Separated legally separated or otherwise absent from spouse due to
marital discord
    Never Married single and never married

Cohabitor (Unmarried Partner) A not currently married adult who shares living quarters and has a close personal relationship with another adult.

Children Ever Born Based on response to a census question asked to women ages 15 and over, married or unmarried. Includes all children born to a woman, including those no longer alive or no longer living with the mother. When compiled for women who are past their prime childbearing ages, this statistic can be used to measure the completed childbearing of these women.

OTHER *concepts*

Race/Ethnicity (Topic two)   **Labor Force Status** (Topic four)
Education (Topic two)   **Cohort** (Topic one)

A. Marital Trends

While you know that the high marriage and birth rates of the 50s dropped off in later decades, it is important to examine how these trends differ by race/ethnicity, age, and education level. Are people more likely to never get married, or are they simply putting off marriage? Is a person's educational attainment related to his or her marital status and/or the age at which he or she marries?

In the following exercises you will look at changes in marital status distribution over time, race/ethnic differences in marital trends, the shift in the average age of marriage, and the relationship between educational attainment and marriage.

Exercise 1 Examine the marital status distribution of Americans from 1950 to 1990. What types of patterns do you see? (*MARR5090.DAT*)

■ Create a stacked bar chart; for each year, stack by marital status.

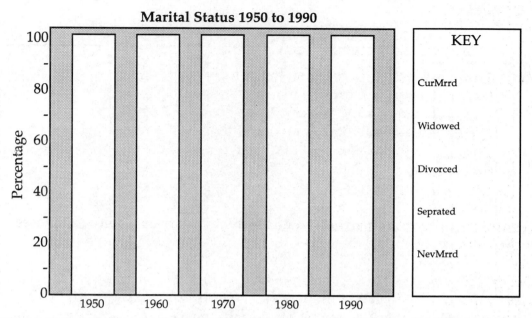

Marital Status 1950 to 1990

Exercise 2 Using data from 1950 to 1990, look specifically at the marital status distribution of people ages 15-24. What trends do you notice? *(MARR5090.DAT)*

■ Create a line graph with three lines, one for currently married, one for divorced, and one for never married. For each year, indicate the percentage of 15-24 year olds in each marital category.

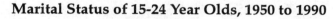

Marital Status of 15-24 Year Olds, 1950 to 1990

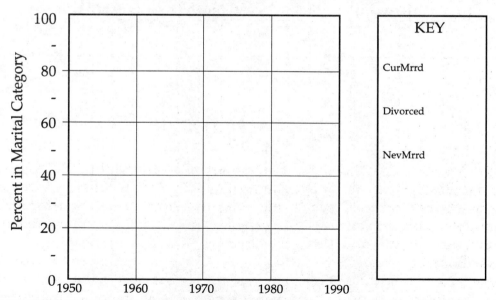

Exercise 3 In the past, it was expected that people would marry when they were 18-21 years old. Has that changed? Using 1990 data, show the marital status distribution of each age group. What is the dominant marital status in each age group? Do your findings surprise you? *(MARITAL9.DAT)*

■ Create a stacked bar chart with bars for each age group. For each age group, stack by marital status.

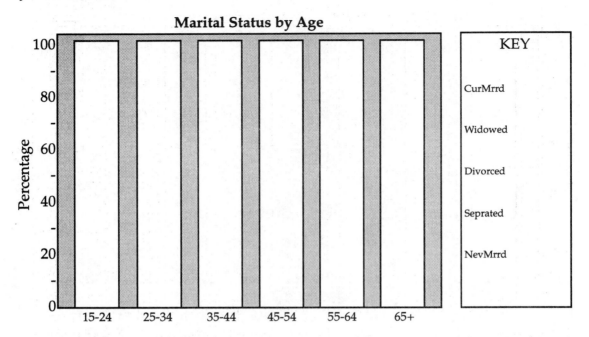

Marital Status by Age

Exercise 4 Examine the marital status differences between blacks and nonblacks from 1950 to 1990. Discuss your findings. *(MARR5090.DAT)*

■ Create a stacked bar chart with side by side bars for blacks and nonblacks. For each year, stack by marital status.

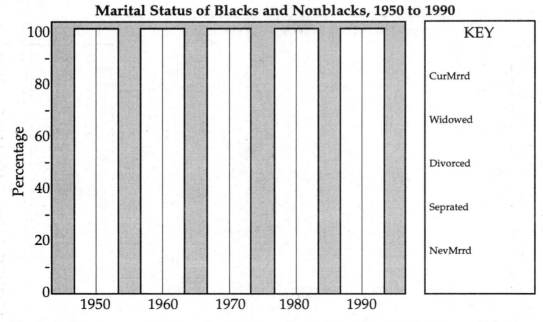

Marital Status of Blacks and Nonblacks, 1950 to 1990

Exercise 5 Focusing on 1990, compare the marital status distribution of blacks, Latinos, whites, Asians, and American Indians. Describe any significant findings. *(MARITAL9.DAT)*

■ Create a stacked bar chart with a bar for each race/ethnic group; stack by marital status.

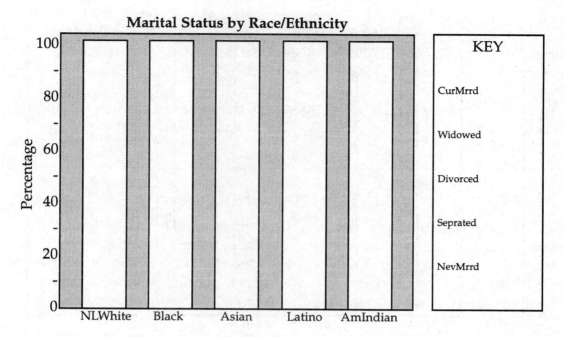

Marital Status by Race/Ethnicity

Exercise 6 Using 1990 data, examine the education levels of 23-28 year old women who have never been married. Describe any significant findings. (*MRR9-YW.DAT*)

■ Create a bar chart with side by side bars for high school and college graduates. For each age, show the percentage of those who have never been married in each education level.

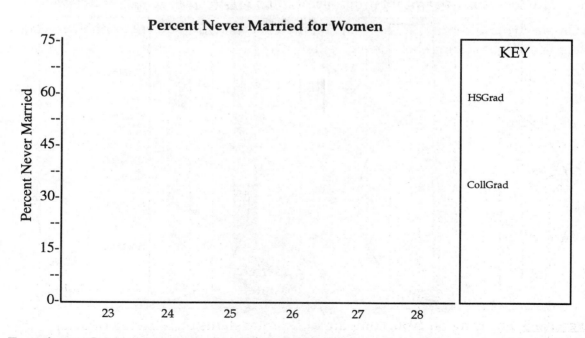

Percent Never Married for Women

Exercise 7 On your own, repeat the previous exercise for women of another race/ ethnicity. (*MRR9-YW.DAT*)

𝒟iscussion Questions

1. Focusing on Americans who have never been married, describe the trends in this marital category since 1950. Highlight any differences between genders and age groups. Would you claim that changes in the percentage of people never married is more significant in certain age groups?

2. How are patterns of marital status specific to different age groups? Do marital trends seem more strongly related to decades or age groups? How might you explain these trends?

𝓑. Marriage Choices

As Americans' attitudes about age, race/ethnicity and education have changed, so have their marriage preferences. While the term "intermarriage" is often applied to the marriage of two people of different race/ethnic backgrounds, race/ethnicity is not the only demographic characteristic which may vary between two spouses. Age and educational attainment can play equally important roles in marriage choices.

In the following exercises, you will look at marriage choices in terms of age, race/ethnicity and education. While completing these exercises, you should consider how spouses have become more different from each other, and how they have become more similar.

Exercise 8 Compare the age distribution of married women to that of married men. Are there any significant differences? *(MARITAL9.DAT)*

■ Create two pie charts, one for married men and one for married women. In each pie, make divisions for each age category.

Age Distributions

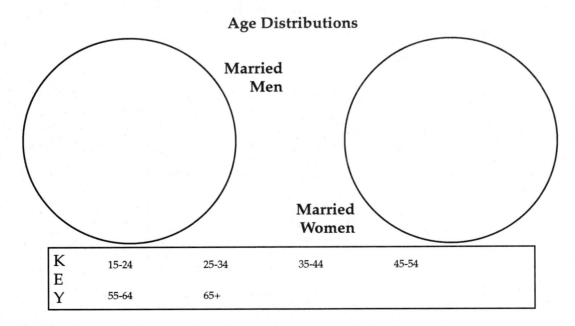

Married
Men

Married
Women

| K E Y | 15-24 | 25-34 | 35-44 | 45-54 |
| | 55-64 | 65+ | | |

Exercise 9 Look at the ages of the spouses of 25 year old men and women. Do men tend to marry younger or older women? How about women? (*SPAGE9YM.DAT, SPAGE9YW.DAT*)

■ Create a stacked bar chart with bars for 25 year old men and 25 year old women showing the age distribution of their spouses.

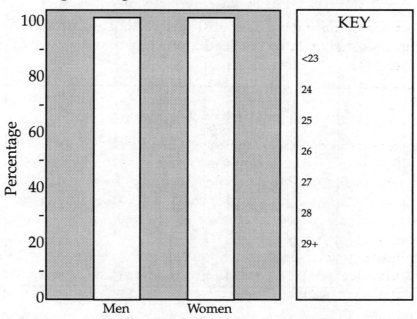

Spouse Ages of 25 Year Old Men and Women

Exercise 10 What proportion of white men are choosing to marry an individual of a different race/ethnicity? How does the percentage of white men who intermarry vary by age group? (*SPRAC9-M.DAT*)

■ Create a bar chart. For each age group, indicate the percentage of white men who are intermarried.

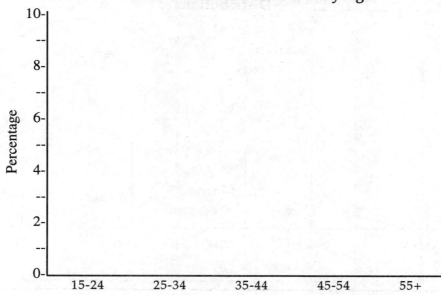

Percent Intermarried for Men by Age

Exercise 11 Look at the race/ethnicity distribution of husbands of white, black, Asian, and Latina women, ages 25-34 in 1990. What differences between race/ethnic groups do you note? *(SPRAC9-W.DAT)*

■ Create a stacked bar chart with bars for each race/ethnic group of wives ages 25-34; stack by the race/ethnicity of the husbands.

Race/Ethnicity of Husbands by Race/Ethnicity of Wives Ages 25-34

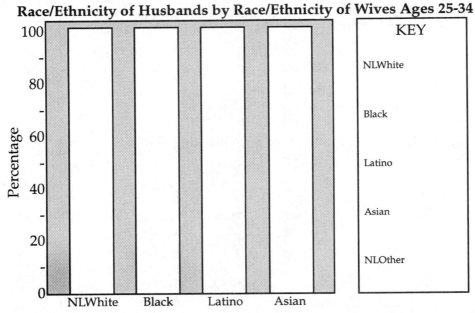

Exercise 12 In addition to age and race, education plays a role in marriage choices. Focusing on women ages 25-34 in 1990, look at the educational attainment distribution of their husbands. Do women tend to marry men who have similar educational backgrounds? *(SPED9-W.DAT)*

■ Create a stacked bar chart; for each education level of wives ages 25-34, stack by the husbands' education levels.

Education of Husbands by Education of Wives Ages 25-34

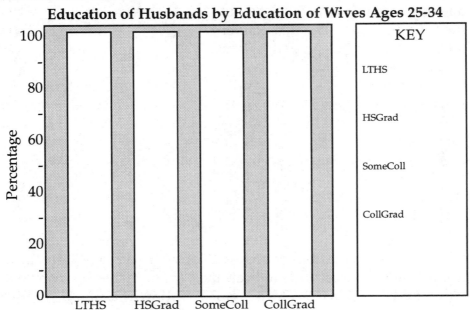

Exercise 13 According to 1990 data, what are the most likely demographics of the husband of a 25-34 year old black woman with a college degree or more? What are the most likely demographics of the husband of a 25-34 year old Latina woman with a college degree or more? On your own, compare the marriage choices of Latina and black women, ages 25-34, with college degrees. (Hint: you must look at race/ethnicity and education separately). (*SPRAC9-W.DAT, SPED9-W.DAT*)

𝒟iscussion Questions

1. Why do you think there are differences in the age distribution of married men and women?

2. Which historical events may have influenced the increase in interracial couples and marriages? By looking at marriage patterns, determine which age group seems to have been most affected by these events.

3. Imagine yourself five years from now. Given your age, race, gender, and expected educational attainment, what are the likely demographics of your future spouse (if you were to marry)? How does this compare to what you would expect?

4. How might you explain recent trends in marriage choices? How do the preferences of certain groups differ from those of their elders? What factors may have influenced these trends and/or changes in people's preferences regarding the age, race/ethnicity, and educational attainment of their spouses?

C. Divorce

In order to obtain a divorce before 1969, one had to prove that his or her spouse was guilty of adultery, desertion, physical or mental abuse, habitual drunkenness or a felony conviction. In 1969, the divorce laws were changed, allowing people to file for a "no-fault" divorce. Rather than placing blame on one particular spouse, a no-fault divorce acknowledges mutual responsibility for the dissolution of the marriage.

While many debate whether no-fault divorce is responsible for the increased divorce rates, one thing is certain: divorce rates rose dramatically between the 1960s and 1970s. It is useful to look at who is getting divorced. Does the likelihood of divorce rates vary between race/ethnic groups? Age groups? Is someone with a high school diploma more likely to get divorced than someone with a Ph.D? Is there a relationship between divorce and labor force participation among women?

In the following exercises, you will look at trends in divorce in terms of age, race/ethnicity, education and labor force participation.

Exercise 14 Examine the percentage of people divorced in each age group in 1990. Do the older age groups' patterns necessarily indicate the percentage of people in your age group who will be divorced as you become those ages? Why or why not? (*MARITAL9.DAT*)

■ Create a bar chart; for each age group, indicate the percentage of those divorced.

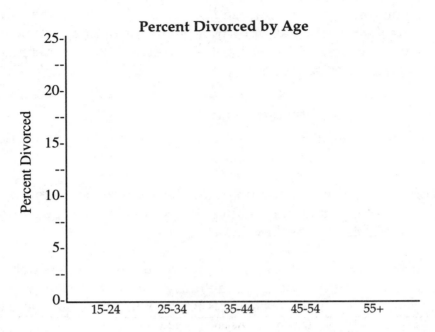

Percent Divorced by Age

Exercise 15 Using data from 1950 to 1990, analyze the differences in the proportion of divorced people among blacks and nonblacks. Discuss significant differences between races and changes over time. (*MARR5090.DAT*)

■ Create a line graph with separate lines for blacks and nonblacks. For each year, indicate the percentage divorced.

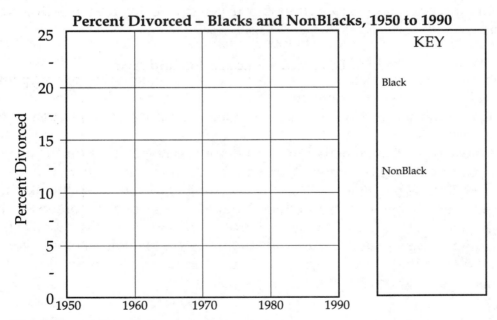

Percent Divorced – Blacks and NonBlacks, 1950 to 1990

KEY

Black

NonBlack

Exercise 16 Combine 15-54 year olds into one age group in order to compare the percentages of people divorced and separated in each race/ ethnic group in 1990. Why do you think these differences exist? (*MARITAL9.DAT*)

■ Create a bar chart with side by side bars for divorce and separation. For each race/ethnic group, indicate the percentage of people divorced and separated.

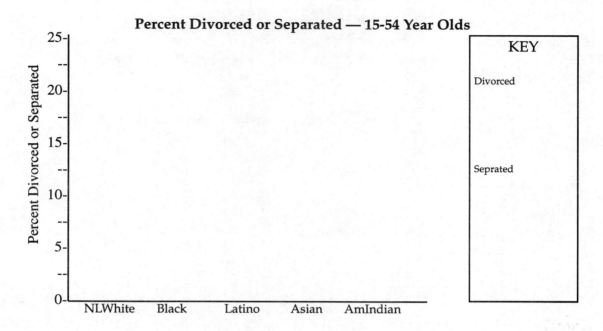

Percent Divorced or Separated — 15-54 Year Olds

Exercise 17 To what extent does there seem to be a connection between educational attainment and divorce? Do people with higher levels of education seem more likely to get divorced? (*MARED9.DAT*)

■ Create a bar chart with side by side bars for three education levels (LTHS, HSGrad, and CollGrad). For each age group, indicate the percentage divorced in each education level.

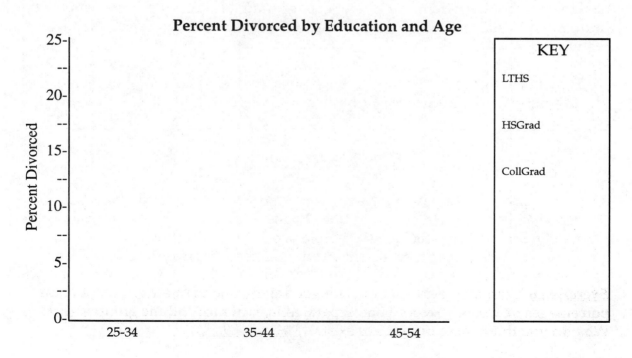

Percent Divorced by Education and Age

Exercise 18 Focusing on women ages 25-34 in 1990, show the relationship between marital status and labor force participation. Is there a greater percentage of those who work outside the home among divorced women than married women? Keeping in mind the patterns you saw in the labor force chapter, describe and offer possible explanations for your findings. (*MREMPF9.DAT*)

■ Create a bar chart; for each marital status, indicate the percentage of women in the labor force. (Hint: Combine EmpFull, EmpPart, and Unempd. See "Key concepts" in Labor Force chapter.)

Percent in Labor Force – Women Ages 25-34

𝒟iscussion Questions

1. How might recent trends in gender equality and the women's movement have influenced women's labor force participation, marital status, and the relationship between the two?

2. Explain how gender relations, race relations, economic trends, labor force participation rates and changing attitudes may affect divorce rates.

𝒟. Cohabitation

The rise in cohabitation rates seems to be concurrent with the maturation of the baby boomers. But who is cohabiting? Young college students? Divorced people who don't want to get remarried for awhile, if ever? Is cohabiting a way to "try out" a relationship before marriage, or is it an alternative to marriage?

In the following exercises, you will examine the characteristics of cohabitants. By looking at the age, race/ethnicity, education, and marital status of cohabitants, you will gain a clearer picture of the role of cohabitation in today's family structure trends.

Exercise 19 Examine the age distribution of people who are currently cohabiting. Do your findings support your expectations? Are there any notable differences between the age distributions of men and women cohabiting? *(COHAB9-M.DAT, COHAB9-W.DAT)*

■ Create two pie charts, one for each gender. In each pie, make divisions for age groups.

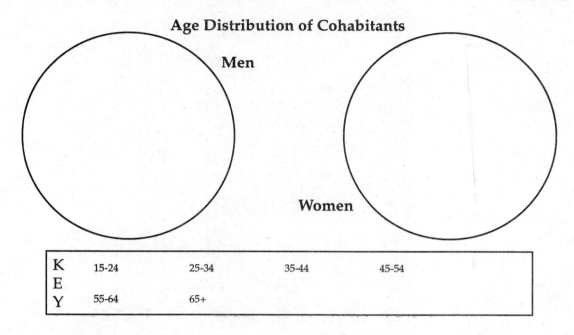

Age Distribution of Cohabitants

Men

Women

| K
E
Y | 15-24 | 25-34 | 35-44 | 45-54 |
| | 55-64 | 65+ | | |

Exercise 20 Does the age distribution of cohabitants in 1990 differ by race? Look at the age distribution of women cohabitants in each race/ ethnic group. Are there any significant differences? Give possible explanations about why some groups may be more or less likely to cohabit. *(COHAB9-W.DAT)*

■ Create a stacked bar chart; for each race/ethnic group, stack by age of cohabitants.

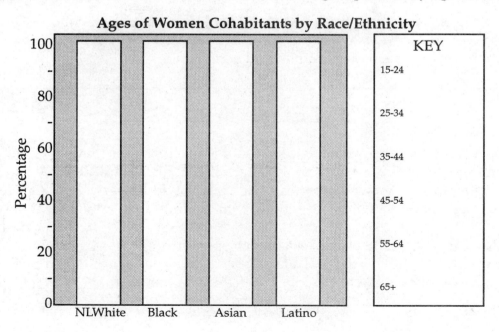

Ages of Women Cohabitants by Race/Ethnicity

KEY

15-24

25-34

35-44

45-54

55-64

65+

Exercise 21 When many people think about cohabitation, they picture college students or recent graduates living together for a short period of time before marriage. Examine and describe the relationship between educational attainment and cohabitation for women. *(COHAB9-W.DAT)*

■ Create a pie chart for women cohabitants in 1990; make divisions for education levels.

Educational Distribution of Women Cohabitants

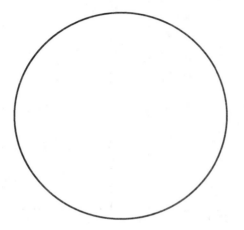 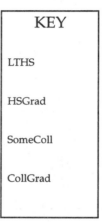

KEY

LTHS

HSGrad

SomeColl

CollGrad

𝒟iscussion Questions

1. Describe the characteristics of cohabiting Americans. Who is most likely to be cohabiting? Offer possible explanations for your findings.

2. Do you think that you are likely to cohabit? Under what circumstances? Did or are either of your parents or grandparents cohabit(ing)? Why do you think cohabitation has become more common and acceptable?

𝓔. Childbearing

In previous sections, we mentioned that the decline in marriage rates and increase in divorce rates is concurrent with a decline in birth rates. To what extent has the number of women who have ever had children declined over the past few decades? Is this trend related to race/ethnicity and educational attainment?

Exercise 22 Using 1950 to 1990 data, examine the percentage of 35-44 year old women who have ever had children. Also look at the percentage of 35-44 year old women who have had three or more children. Describe any significant trends. *(BORN5090.DAT)*

■ Create a line graph with two lines, one for women who have ever had one or more children and one for women who have had three or more children. For each year, indicate the percentages.

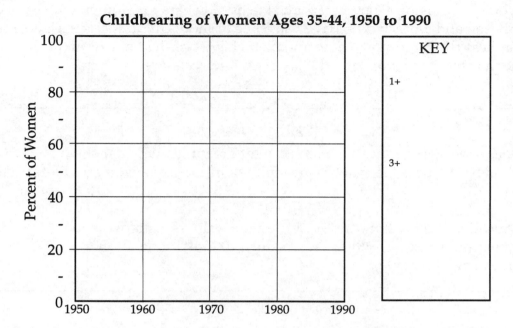

Childbearing of Women Ages 35-44, 1950 to 1990

Exercise 23 Now focus on 1990. For each race/ethnic group, look at the percentage of women, ages 35-44, who have ever had one or more children and those who have had three or more children. Are there differences between race/ethnic groups? (*BORN9.DAT*)

■ Create a bar chart with side by side bars for women who have ever had children and women who have had three or more children. For each race/ ethnic group, indicate the percentage of women who have ever had one or more children and the percentage of women who have had three or more children.

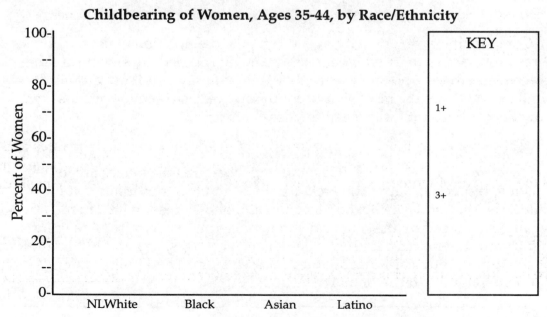

Childbearing of Women, Ages 35-44, by Race/Ethnicity

Exercise 24 Does a woman's education level play a role in her childbearing decisions? Using 1990 data, determine the relationship between education and childbearing for women ages 35-44. *(BORN9.DAT)*

■ On your own, create a bar chart with side by side bars for women who have ever had children and women who have had three or more children. For each education level, indicate the percentage of women who have ever had one or more children and the percentage of women who have had three or more children.

---

*D*iscussion Questions

1. Does a woman's educational attainment determine whether she will have children or is it possible that having children influences a woman's education level?

2. Education, occupation, and society's mores all influence women's childbearing decisions. Which factors do you think influence men's childbearing decisions?

THINK*tank*

1. Compare the race/ethnicity, age, poverty, and employment characteristics of three groups of unmarried women: those never married, those who are divorced, and those who are widowed. Would you say that these groups are similar? What are the most significant differences between the groups, and how do you account for these differences?

2. Some social critics suggest that the *form* of marriage is less important than the *function*. In other words, it is more important that marriages are loving, trusting, respectful, supportive relationships, and less important that they reflect the traditional context of one man and one woman in a relationship legally recognized by the state. Do you agree with this premise? Are there significant differences between male-female marriages, gay and lesbian unions, or cohabiting couples? What are the social, economic, and political implications of broadening the 'form' of marriage in the U.S.?

The image of the "typical" American woman has changed dramatically since the 1950s. As the number of women graduating from college and joining the labor force steadily increases, there has been a significant decrease in the number of women who choose to devote all of their energies to raising a family. Yet, personal choices are not always as clear cut as the overall trends may indicate. Many women are still grappling with how to balance their personal and professional lives.

The changing role of women in society is strongly related to historical events and the economy. While working class women have always worked outside the home, the emergence of all classes of women into the workplace can be linked to World War II when women worked while men were at war. During this period, many women enjoyed their new independence and responsibilities, but most were forced to leave their positions when the men came back home. Although opportunities for women to work outside the home were limited during the 50s, the strong economy and increased federal support enabled many women in the baby boom cohort to attend college in the 1960s. Women were more likely to delay marriage to pursue careers. Since 1970, women's occupational choices have been affected by the economy and industrial shifts from manufacturing to services. During recent decades, many families have found that they cannot survive on the earnings of one breadwinner. The need for a second salary has not only had a significant impact on women, but has also changed society's perceptions of women.

Despite many gains, women are still not equally represented in all fields, do not advance at the same rate as men do, and are not equally represented in all income brackets. In order to understand gender inequality in the workplace, you will examine the earnings and occupational choices of men and women with the same level of education. Furthermore, you can consider whether the gender earnings gap result from discriminatory practices such as the "glass ceiling" and the "old boy network," or from factors such as real experience differences between men and women.

By charting trends in women's educational attainment, occupational choices and earnings, you will consider how, why, and to what extent women's lives have changed since 1950.

KEY concepts

Gender *Male* or *Female*

OTHER concepts

Education (Topic two)   **Race/Ethnicity** (Topic two)
Occupation (Topic two)   **Children Ever Born** (Topic five)
Earnings (Topic two)   **Cohort** (Topic one)

A. Education and Gender

Until the 1950s, the cost of a college education limited enrollment to mostly students from wealthy families. Often, if a family could afford the tuition for only one child, they would choose to send their son. An abundance of jobs and higher salaries in the 1950s and 1960s enabled more families to send their children to college. Beginning in the 1960s, increased government support for higher education increased opportunities for both women and men. Since then, women, as well as previously underrepresented race/ethnic groups, have begun to attend college in much larger numbers.

In addition to considering economic trends, many people believe the educational attainment of women is linked to childbearing trends. During the past few decades, as their education levels have increased, women have been having fewer children, and having them later in life.

In the following exercises, you will examine trends in educational attainment since 1950, including differences by gender and gender gaps within different race/ethnic groups. You will also look at the relationship between childbearing and educational attainment.

Exercise 1 Using data from 1950 to 1990, examine changes in the percentage of men and women ages 25-34 with a high school education or more. Describe the role gender seems to play in educational attainment, and how the gender gap may have changed over time. Why do you think it is useful to focus on this particular age group? (*EDUC5090.DAT*)

■ Create a line graph with a line for women and a line for men. For each year, indicate the percentage of 25-34 year old men and women with a high school education or more. (Hint: Combine high school graduates, those with some college, and college graduates.)

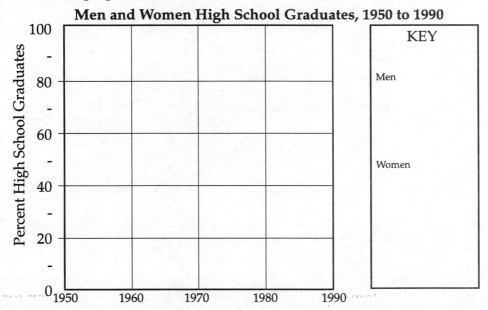

Exercise 2 Using data from 1950 and 1990, look at the difference between the percentage of high school graduates among black and nonblack men and women, ages 25-34. Describe any significant differences. (*EDUC5090.DAT*)

■ Create two bar charts, one for blacks and one for nonblacks, with side by side bars for men and women. In each chart, for both 1950 and 1990, show the percentage of 25-34 year old men and women with a high school education or more. (Hint: Combine high school graduates, those with some college, and college graduates.)

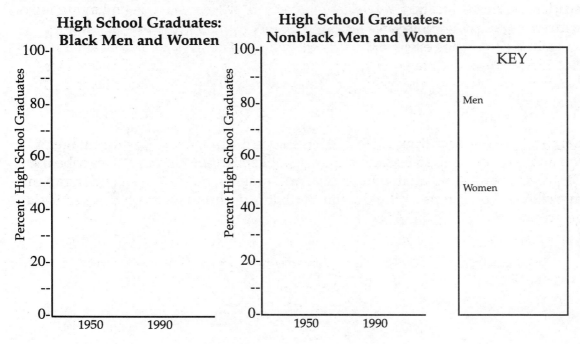

Exercise 3 Using data from 1950 to 1990, examine changes in the percentages of men and women, ages 25-34, who have graduated from college. Is the gap between men and women at this level of educational attainment significantly different from the gender gap at the high school diploma level? *(EDUC5090.DAT)*

■ Create a line graph with a line for women and a line for men, ages 25-34. For each year, indicate the percentage of college graduates in each group.

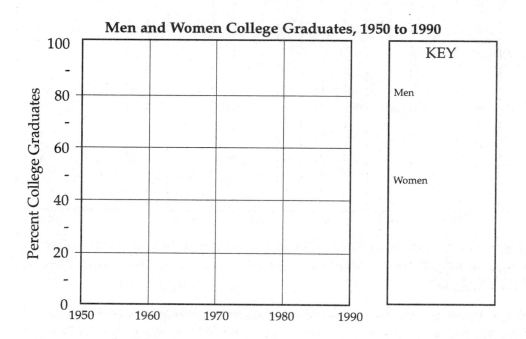

Men and Women College Graduates, 1950 to 1990

Exercise 4 What are the differences in college degree attainment between black and nonblack men and women, ages 25-34, from 1950 to 1990? Are the gender and race/ethnicity gaps between these four groups significantly different from the gaps at the high school diploma level? *(EDUC5090.DAT)*

■ On your own, create two bar charts, one for blacks and one for nonblacks, with side by side bars for men and women. In each chart, for each year, show the percentage of 25-34 year old men and women with a college degree or more.

Exercise 5 On your own, examine the educational attainment of women ages 25 and above in 1990. Describe the differences between the younger and older cohorts. Which cohorts of women appear to have made the greatest gains in educational attainment? *(EDUCIMM9)*

Exercise 6 Focusing on 1990, show the educational attainment distribution of men and women, ages 25-34, in all race/ethnic groups. How does the gender gap differ between race/ethnic groups? *(EDUCIMM9.DAT)*

■ Create a stacked bar chart with side by side bars for men and women, ages 25-34. For each race/ethnic group, stack by educational attainment.

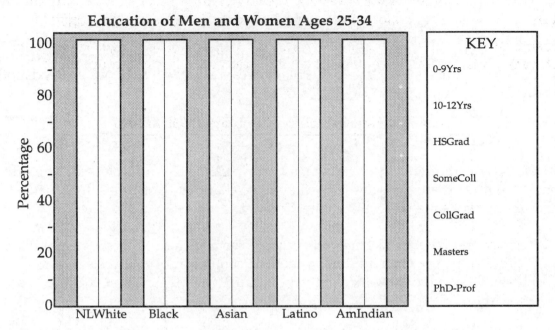

Education of Men and Women Ages 25-34

KEY
0-9Yrs
10-12Yrs
HSGrad
SomeColl
CollGrad
Masters
PhD-Prof

Exercise 7 On your own, focus on people with M.A.'s and Ph.D's and describe the gender gap among those with higher levels of educational attainment in 1990. (*EDUCIMM9.DAT*)

Exercise 8 Looking at women ages 45-54, illustrate the relationship between childbearing and women's educational attainment in 1990. (*BORN9.DAT*)

■ Create a stacked bar chart with bars for each education level; stack by the number of children.

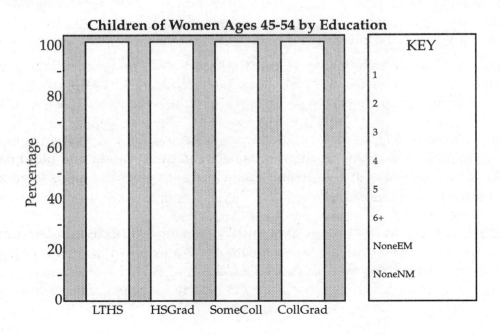

Children of Women Ages 45-54 by Education

KEY
1
2
3
4
5
6+
NoneEM
NoneNM

Exercise 9 On your own, focus on women with the highest levels of educational attainment. Explain how their childbearing experiences are different from those of women with fewer years of education in 1990. (*BORN9.DAT*)

Exercise 10 How does looking at race/ethnicity affect the results of exercises 8 and 9? On your own, examine the relationship of educational attainment and childbearing in two race/ethnic groups of your choice. Describe the differences you note between overall trends and individual race/ethnic groups. (*BORN9.DAT*)

---

*D*iscussion **Questions**

1. Discuss the differences between men and women's education levels. How do you account for these gaps and why they have changed over time? In which areas are the gaps the largest? Smallest?

2. Which factors do you think have contributed to women's increased educational attainment?   Focusing on women in 1990, what predictions would you make for the educational attainment of women in the year 2010? On which factors did you base your prediction?

3. Looking at the different education levels of Latina, black and Asian women in 1990, offer some possible causes of these differences.

4. How does your level of educational attainment compare to your parents and grandparents' educations? Which factors may have affected these generational differences?

5. Why do women with higher levels of education in 1990 appear to be putting off having children while they finish their education?

B. Occupation and Gender

Historically, there has been a very clear division of labor between men and women. Until recently, most women entered the fields of nursing, social work, teaching, service and domestic work. Now, women are moving into all fields, and the number of male and female professionals entering the labor force is more equal. Conversely, men are slowly moving into fields traditionally dominated by women. Even as men and women begin to enter some fields at a more equal rate, internal gender hierarchies are still evident within occupations.

Many believe that a "double standard" affects the role of women in the workplace, even in upper management positions. For example, a man striving to climb the corporate ladder may be viewed as "ambitious," while his equally driven female counterpart may be considered "pushy." Women, more often than men, are still sexually harassed in the workplace.

Despite the growing popularity of flex time, day care and fathers' increased contributions to their family responsibilities, many women have not been able to devote all of their energies to their careers due to their roles as their families' primary caretakers. In the following exercises, you will examine trends in occupational choices among men and women. You will also look at these gender divisions among different race/ethnic groups.

Exercise 11 Using data from 1950 and 1990, look at the changing percentages of women and men, ages 35-44, in each occupational category. Do men and women appear to be concentrated in different occupational categories? How has the distribution changed over time? (*EDOC5090.DAT*)

■ Create two bar charts, one for each year. Using side by side bars for men and women, ages 35-44, indicate the percentage in each occupational category.

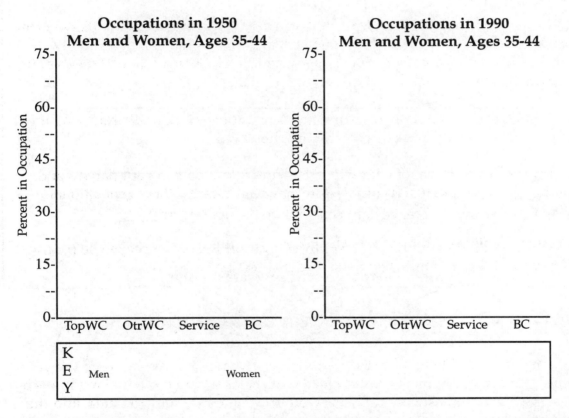

Exercise 12 How does the occupational distribution vary for black and nonblack men and women, ages 35-44, over time? In which occupations are black men and women more likely to work? On your own, examine gender differences within each race and then compare your findings to the gender gaps in the entire population (exercise 11). (*EDOC5090.DAT*)

Exercise 13 Investigate 1990 gender differences in occupations for specific Asian groups, ages 35-44. Describe your findings and compare them to the gender gaps of other race/ethnic groups. (*OCCASN9.DAT*)

■ On your own, create a stacked bar chart with side by side bars for women and men, ages 35-44 in each specific Asian group; stack by occupational category.

Exercise 14 Compare the gender distribution of doctors ages 25-34 to doctors ages 55-64. Is there any difference between the two groups? (*DOCTORS9.DAT*)

■ Create two pie charts, one for each age group, with divisions for gender.

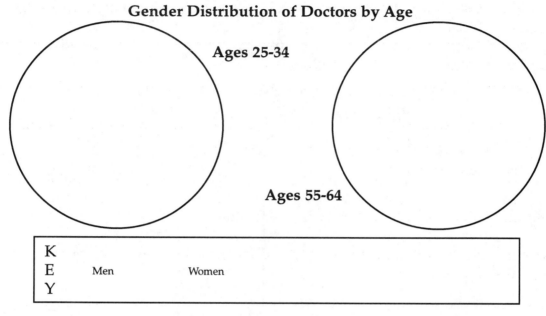

Gender Distribution of Doctors by Age

Ages 25-34

Ages 55-64

K
E   Men   Women
Y

Exercise 15 Compare the gender distribution of lawyers ages 25-34 to lawyers ages 55-64. Is there any difference between the two groups? (*LAWYERS9.DAT*)

■ Create two pie charts, one for each age group, with divisions for gender.

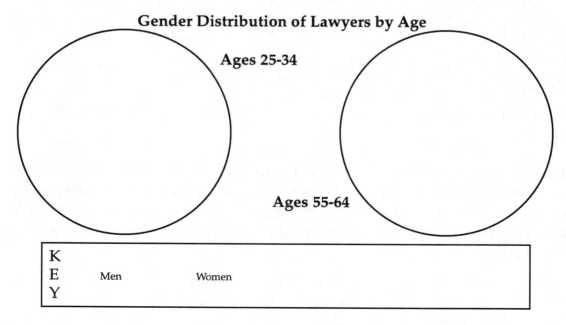

Gender Distribution of Lawyers by Age

Ages 25-34

Ages 55-64

K
E   Men   Women
Y

Exercise 16 Look at the gender distribution of doctors and lawyers, ages 25-34 and 55-64, in each race/ethnic group. Compare these findings to the gender distributions in exercises 14 and 15. (*DOCTORS9.DAT, LAWYERS9.DAT*)

■ On your own, create a stacked bar chart showing the gender distribution of doctors in each race/ethnic group. Use side by side bars for each age group; stack by gender.

■ On your own, create a stacked bar chart showing the gender distribution of lawyers in each race/ethnic group. Use side by side bars for each age group; stack by gender.

---

*D*iscussion Questions

1. What factors affect a woman's occupational choice? A man's?

2. How does society perceive occupations traditionally held by women? Men? Do you think that society values certain occupations more than others?

3. Why do you think that a great number of women have moved into fields previously dominated by men, while a relatively small number of men have moved into fields previously dominated by women?

4. Why do you think that men are disproportionately represented among the higher ranks in fields dominated by women, such as nursing?

C. Education and Occupation

An individual's educational attainment strongly affects his or her occupational choice and status. As we discussed in the previous section, women and men have tended to enter different fields until recently. Similarly, they have chosen different areas of study and have not pursued higher degrees at equal rates. Although the gender differences in educational attainment and occupational choice are becoming narrower, they still exist in many situations. In the following exercises, you will explore how educational attainment affects men and women's occupational choices.

Exercise 17 Using data from 1970 and 1990, examine how educational attainment affects the occupational choices of working men and women, ages 35-44. Compare the occupations of women and men who are college graduates and describe your findings. Has there been any change over time? Why is it helpful to focus on 1970 and 1990? (*EDOC5090.DAT*)

■ Create a stacked bar chart with side by side bars for male and female college graduates, ages 35-44; for each year, stack by occupation.

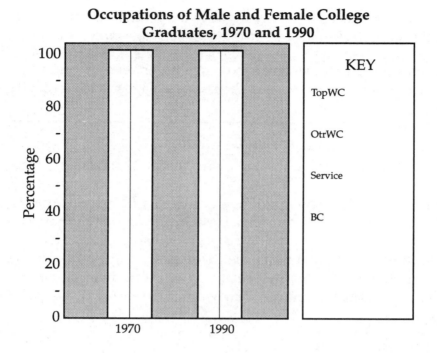

Occupations of Male and Female College Graduates, 1970 and 1990

KEY

TopWC

OtrWC

Service

BC

Exercise 18 On your own, look at the data in exercise 17 in terms of race/ethnicity. Are there gender differences within and between race/ethnic groups? Describe your findings. (*EDOC5090.DAT*)

---

*D*iscussion Questions

1. How do you think your gender has affected your study and career plans?

2. How do your career plans compare to your parents and grandparents' occupational choices? Which factors may have affected these generational differences?

3. How do you think the level of educational attainment affects women and men's occupational choices differently?

4. In the future, do you think the occupational choices of men and women with comparable educational backgrounds will become more similar?

D. The Gender Gap in Earnings

Over the last few decades, women have earned more money as their level of educational attainment and occupational status increases. However, a gap between male and female earnings persists in most occupations. On an individual level, a woman often earns less than a man who has an identical position in the same company. Sometimes, these differences are attributed to various forms of gender discrimination which may be fostered by the "glass ceiling" or the "old boy network." However, as shown in the Labor Force chapter, women with seemingly similar credentials sometimes have fewer years of full-time work experience than men because they have left the labor force due to family or other reasons. Making comparisons among equally qualified groups is difficult even for social scientists.

In the following exercises, you will compare the earnings of men and women who work full-time, year-round. You will take a closer look at the earnings gender gap in the 1990s by comparing men and women with similar education levels and occupations.

Exercise 19 Using 1990 data, compare the earnings of women and men ages 25-34 to those ages 35-44. Describe and account for any differences. (*EARN9.DAT*)

■ Create a stacked bar chart with side by side bars for men and women. Draw a bar for 25-34 year olds and a bar for 35-44 year olds; stack by 1990 earnings.

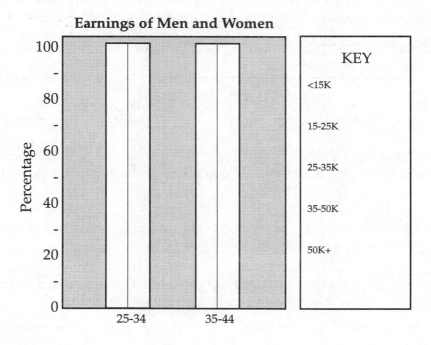

Exercise 20 Look at the earnings of men and women, ages 35-44, in each race/ethnic group in 1990. What are the gender differences within and between race/ethnic groups? (*EARN9.DAT*)

■ Create a stacked bar chart with side by side bars for men and women; for each race/ethnic group, stack by earnings.

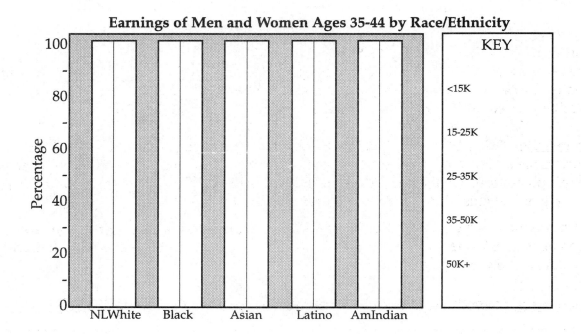

Earnings of Men and Women Ages 35-44 by Race/Ethnicity

Exercise 21 Looking at the earnings of men and women, ages 35-44, in 1990, focus on education. Does educational attainment affect the gender gaps you saw in the previous exercises? (*WORK9-35.DAT*)

■ Create a stacked bar chart with side by side bars for men and women; for each education level, stack by earnings.

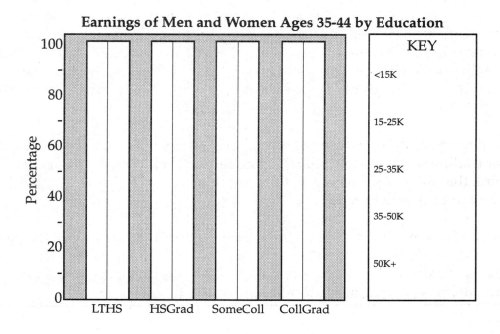

Earnings of Men and Women Ages 35-44 by Education

Exercise 22 On your own, determine whether a college education decreases the gender gap in earnings for 35-44 year olds in all race/ethnic groups. If not, why do you think some differences persist? (*WORK9-35.DAT, WKLT9-35.DAT, WKAS9-35.DAT*)

Exercise 23 Is it possible to erase the gender gap by focusing on those who not only have college degrees, but also top white collar jobs? On your own, examine the gender gap in the earnings of 35-44 year olds with a college education in top white collar jobs. Do you continue to see differences between women and men? Do both genders seem to be receiving an equal "return" for their education levels? (*WORK9-35.DAT*)

Exercise 24 Using 1990 data, examine the gender gap in doctors' earnings. Look at doctors ages 25-34 and 45-64 and compare the differences in women and men's earnings in this profession. Do the patterns seem to be any different from those you found when looking at all fields? How does the gender gap vary by age? (*DOCTORS9.DAT*)

■ On your own, create four pie charts for doctors: one for men ages 25-34, one for men ages 45-64, one for women ages 25-34, and one for women ages 45-64. In each pie, show the earnings distribution.

Exercise 25 On your own, repeat the previous exercise for lawyers. (*LAWYERS9.DAT*)

---

*D*iscussion Questions

1. Why do earning differences between genders exist? Do you think the differences are increasing or diminishing? Why?

2. Why do you think white men are disproportionately represented in the highest earnings categories? How does this affect women and other minorities?

3. How would you explain the fact that fields traditionally dominated by women tend to pay less than other fields? Do you think these fields pay less because they are less valued by society, or do they pay less because women work in them? How do you think the pay would change if these fields were dominated by men?

4. Building upon question three, some women believe that if they convince men to enter traditional "women's fields," more respect and money will follow. Do you think that this would happen? Can you think of any fields in which this has happened or been attempted?

5. Looking back at gender differences in earnings, what do you think the gender distribution will look like twenty years from now? Up to now, how have your earnings compared to those of the opposite sex? How do you think your gender and race affects your earnings potential?

THINK tank

1. Relationships and roles of men and women can sometimes vary a great deal between cultures and race/ethnic groups. Although the U.S. Census isn't designed to directly investigate gender attitudes or expectations, it does offer a great deal of information about the daily lives of men and women. Based on your examination of the data, does it appear that gender differences vary by race/ethnicity? What evidence can you produce to support or disprove the notion that gender is perceived differently by different cultures.

2. The relatively low number of women in high level corporate positions has sometimes been attributed to the *glass ceiling* — a restricted access to upper-level positions because of gender. However, some critics suggest that these differences are not due to biased workplace policies, but rather due to the disruptive effects of childbearing and other family responsibilities. Is this a valid argument? What evidence can you produce to support or disprove this premise? What policies could businesses implement to help women and men balance career and family? Would such policies be good for business? Why or why not?

HOUSEHOLDS AND FAMILIES:
topic seven

This topic is somewhat different from the previous ones because you will not be examining persons as individuals. Instead, you will be looking at groups of individuals who live in households. American households have taken on a variety of forms since the 1950s. The married couple family is still prominent, but most researchers and social observers would agree that households and families have undergone a great deal of change since the days of "Ozzie and Harriet."

While you will be exploring household composition trends from 1950 to 1990, it is important to understand that the "Ozzie and Harriet" households of the 50s were actually somewhat of an anomaly in terms of household history. From 1900 to 1930, the United States experienced a steady decline in fertility, and subsequent shrinking household sizes. The Depression of the 1930s caused marriage rates to plummet and births fell to a record low. Economic uncertainty prompted many couples to delay marriage and child bearing. Furthermore, many singles and families were forced to share housing units. As a result, household formation dropped considerably.

As the nation began its economic recovery in the late 30s and early 40s, marriage, birth and household formation rates started to increase. The beginning of World War II precipitated an economic turnaround, and the conclusion of the war caused a dramatic reversal of demographic trends. Marriage rates surged and America experienced the "Baby Boom." In just twenty years, between 1940 and 1960, the population under age 5 doubled.

In addition to causing an increase in marriage and birth rates, the improved economic situation enabled more people to purchase new single family homes. The number of houses built during the 1950s was double that of the 1940s. This large crop of new homes paved the way for mass movement to suburban America. Small single family homes became the housing standard, and the two parent family with children represented the "typical" household. Although the 1950s may not have been the "golden age" some suggest, it certainly was an era of economic and demographic growth.

By the mid sixties, as the baby boomers matured, marriage and birth rates plummeted and divorce was on the rise. These shifts in marriage, birth and divorce

KEY concepts

Household A household refers to all of the persons (one or more) who occupy a single housing unit. There are two main types of households: "Family Households" and "Nonfamily Households." These are defined below. For census purposes, one person in each household is designated as a "house-holder" (sometimes thought of as the household head). The householder typically is the person, or one of the persons, who owns or rents the home. NOTE: Datasets include variables which pertain to family households only (FAMTYPE3) or all household types, including both family and nonfamily households (HHType4, and HHType5).

Family Household A household in which the head (householder) is related to one or more other person by birth, marriage, or adoption. Family households can be classified as follows:

    Married Couple Family Husband and wife living together along with any other relatives (e.g. children).

    Male-Headed Family (Male Householder Family) A household headed by an adult male, with no spouse present, living with one or more relatives (e.g. single parent families).

    Female-Headed Family (Female Householder Family) A household headed by an adult female, with no spouse present, living with one or more relatives (e.g. single parent families).

Nonfamily Households A household consisting of a single person living alone, or one or more unrelated persons sharing the same housing unit such as housemates, roommates, and boarders. (Unmarried, cohabiting partners are, strictly speaking, included in this category, but are not specifically identified as such.) Nonfamily households can be further classified by the gender of the household head (householder):

    Male Nonfamily Households Male living alone or as household head (householder) living with one or more nonrelatives.

    Female Nonfamily Households Female living alone or as household head (householder) with one or more nonrelatives.

Presence of Children Under 18 Households can be classified by the number of never married own children, under the age of 18, residing in the household. The following classification will be used here:

    No own children present
    At least one child under age 6
    At least one child 6-17 years old, with none under age 6

NOTE: Dataset variable KID3 abbreviates these categories as: None, Kids < 6, and KidsOtr.

Household Size Number of persons per household, with categories _1_, _2_, _3_, _4_, and _5+_.

Dual Earner Family Status Married couple families are classed according to the husband's and wife's labor force status as follows: _Dual Earner_, _Single Earner Male_, _Single Earner Female_, _No Earner Family_.

Housing Type An occupied housing unit is a house, an apartment, a mobile home, a group of rooms, or a single room that is occupied as separate living quarters. Housing type will be classified here as:
>_House_ single family home either detached or attached (town houses)
>_Apartment in building with 2-9 units_
>_Apartment in building with 10+ units_
>_Mobile home or trailer_
>_Other_ includes other categories such as houseboats, campers, vans

NOTE: Dataset variable HOUSING5 abbreviates these categories as House, APT2-9, APT10+, MobHome, and Other.

Ownership-Rentership Classifies households according to the ownership of the housing unit according to the following:
>_Owner_ denotes that the household head (householder) or another household member owns or co-owns the unit.
>_Renter_ denotes all other households

OTHER _concepts_

Race/Ethnicity (Topic two)   **Poverty Status of Family** (Topic eight)

had a significant impact on household composition and formation. The corresponding increase in single parent households and nonfamily households has sparked considerable public debate.

Alternative household arrangements and changing notions of family and marriage challenge current laws regarding adoption, health insurance coverage for unmarried partners, and tax deductions. These changes lead us to question what a "normal" family is.

As you work through the following exercises, think of changes in social attitudes and economic trends, as well as specific historical events, that may have affected households and families. By looking at overall household trends, the characteristics of nonfamily households, married couples, female headed and male headed households, and housing types, you will gain an understanding of the transformation of households since 1950.

𝒜. Household Trends

The size and composition of American households is changing. The two parent standard of the 1950s is slowly giving way to a greater proportion of smaller, nonfamily households and female-headed and male-headed family households. Since 1950, there has been a significant change in the number of households comprised of only one or two people. While a variety of household types existed during the 1950s, the following exercises reveal a trend towards an increasingly complex mosaic of family and nonfamily arrangements.

Exercise 1 During which ten year period did the number of households increase the most? During which ten year period did the total U.S. population increase the most? Using data from 1950 to 1990, look at the changes in the number of households and people in the United States. Why do these trends differ? (*POP5090.DAT, HH5090.DAT*)

■ Create a line graph with two lines, one for the number of households (in millions) and one for the U.S. population (in millions). For each year, indicate the number of households and people living in the United States.

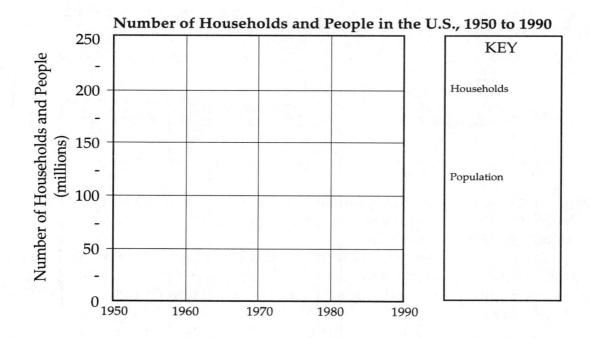

Number of Households and People in the U.S., 1950 to 1990

Exercise 2 Look at the change in household size since 1950. Has there been an overall decrease or increase? How do you account for your findings? (*HH5090.DAT*)

■ Create a stacked bar chart with bars for each year; stack by the percentages of the number of household members.

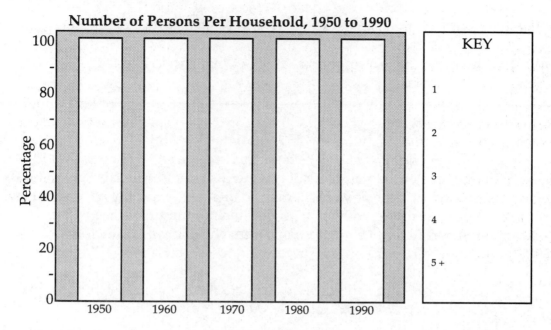

Exercise 3 How has the composition of U.S. households changed over time? Determine the percentage of each household type since 1950 and discuss any significant increases and decreases. *(HH5090.DAT)*

■ Create a stacked bar chart with bars for each year; stack by household type.

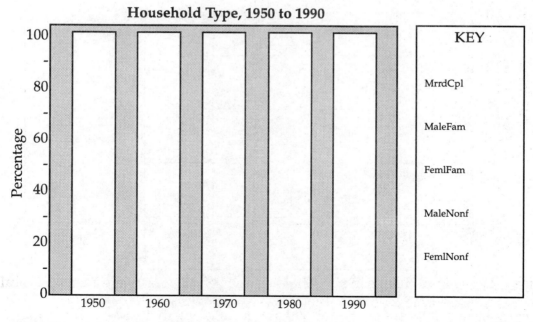

Exercise 4 Using your findings from the previous exercise, describe the changes that occurred for *each* household type over the last four decades. Discuss changes in the different types of family households and nonfamily households. *(HH5090.DAT)*

Discussion Questions

1. Describe a recent historical event or social/economic trend (1970 to 1990) and explain why and how it may have influenced household size or composition.

2. Why has household size changed since 1950? What are the effects of this decline?

$\mathcal{B}$. Nonfamily Households

One of the most significant trends in the last few decades has been a rapid increase in nonfamily households. Most of this increase stems from a growing number of single person households, many of whom are older and financially better-off than in previous eras. These increases, coupled with declining fertility, have been a major contributor to the overall decline in household size. The following exercises look at nonfamily households in terms of gender, age, and race/ethnicity.

Exercise 5 What percentage of households in 1990 were nonfamily households? What has happened to the percentage of nonfamily households over the last forty years? (_HH5090.DAT_)

■ Create a line graph indicating the percentage of nonfamily households from 1950 to 1990.

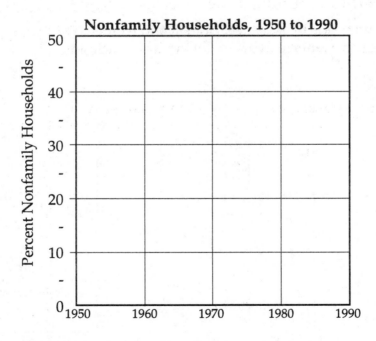

Exercise 6 Using data from 1950 to 1990, determine whether there were as many female nonfamily households as male nonfamily households. What are the trends over time and what might account for the differences? (_HH5090.DAT_)

■ Create a bar chart with side by side bars for male nonfamily and female nonfamily households; for each year, indicate the percentage of all households which are in each of these two categories.

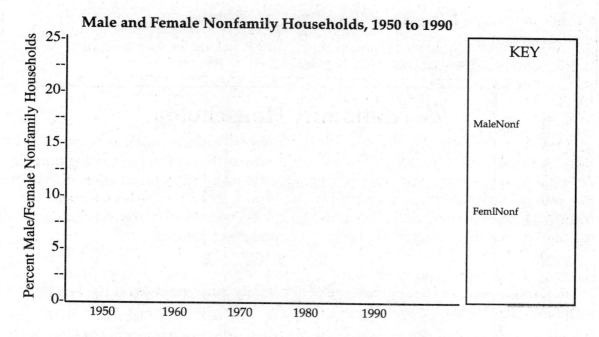

Male and Female Nonfamily Households, 1950 to 1990

Exercise 7 Why might a householder's age affect whether he or she lives in a nonfamily household? How does this differ by gender? *(HHOLDS9.DAT)*

■ Create a bar chart with side by side bars for men and women. For each householder age group, indicate the percentage living in nonfamily households.

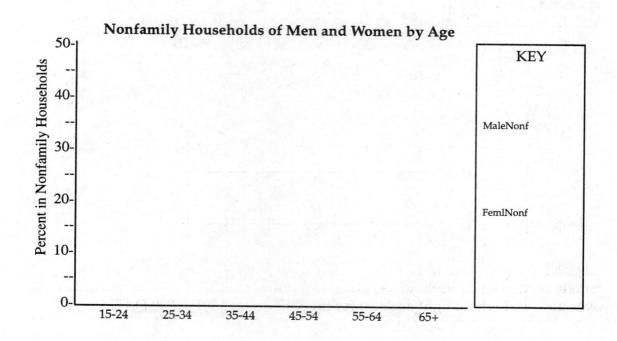

Nonfamily Households of Men and Women by Age

Exercise 8 Using data from 1950 to 1990, show trends in the percentage of male and female nonfamily households for blacks and nonblacks over time. Describe any notable trends. What might account for these trends? (*HH5090.DAT*)

■ Create a line graph indicating the percentage of male nonfamily households and female nonfamily households for blacks and nonblacks, 1950 to 1990.

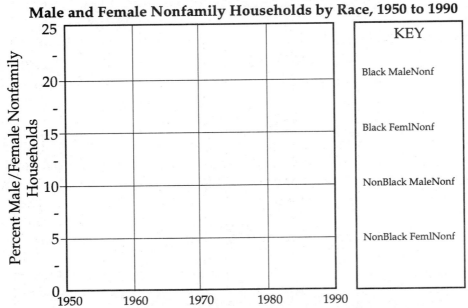

Male and Female Nonfamily Households by Race, 1950 to 1990

Exercise 9 Focusing on 1990, determine the percentage of family households, male nonfamily, and female nonfamily households in each race/ethnic group. Describe any significant differences. (*HHOLDS9.DAT*)

■ Create a stacked bar chart with bars for each race/ethnic group; stack by family, male and female nonfamily households in 1990. (Hint: Combine categories MrrdCpl, MaleFam, and FemlFam for the Family category.)

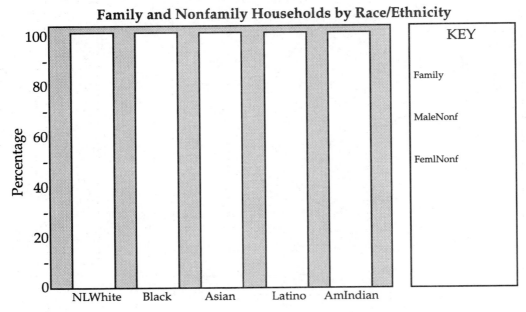

Family and Nonfamily Households by Race/Ethnicity

$\mathcal{D}$iscussion **Questions**

1. Cohabiting couples are currently categorized as nonfamily households. What are the legal and economic implications of not recognizing unmarried couples as families?

2. Why has there been an increase in the number of female nonfamily households?

3. Discuss the potential impact of population aging on nonfamily households. What are some likely effects of a growing number of single elderly people?

C. Married Couples

Despite changes in marriage, childbearing, divorce and cohabitation, the majority of households in the United States are family households. Most of these family households include married couples. A majority of these couples are in their first marriage, but many have been divorced and remarried. While the current trend among young adults is to stay single, demographers estimate that the vast majority of Americans will eventually marry.

<u>**Exercise 10**</u> Considering the changes in the number of married couple households over time (see exercise 3), focus on the family household distribution in 1990. What percentage of family households are married couples, male-headed, and female-headed? (*FAMILY9.DAT*)

■ Create a pie chart with divisions for each type of family household.

Family Household Distribution—1990

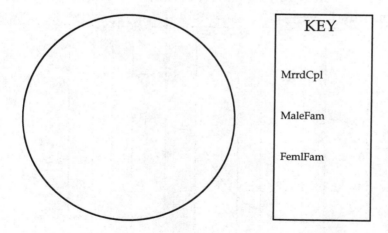

KEY

MrrdCpl

MaleFam

FemlFam

Exercise 11 Using 1990 data, look at the percentage of married couples in each race/ethnic group. Describe any significant findings. (*FAMILY9.DAT*)

■ Create a stacked bar chart with bars for each race/ethnic group; stack by the percentage of married couples and "other" family types in 1990. (Hint: Combine MaleFam with FemlFam.)

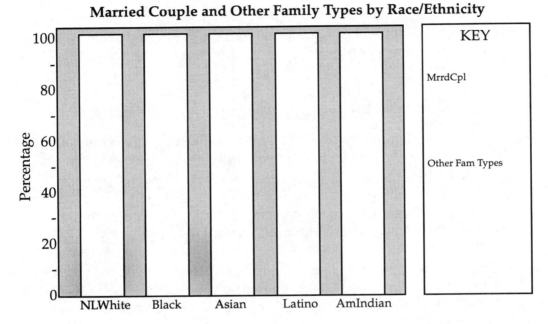

Married Couple and Other Family Types by Race/Ethnicity

Exercise 12 Are children a part of most married couple families in 1990? (*FAMILY9.DAT*)

■ Create a pie chart for married couples with divisions showing the children categories (See "Key Concepts").

Presence of Children in Married Couple Families

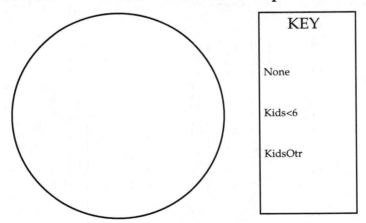

Exercise 13 On your own, using 1990 data, explore the percentage of married couples who have children in each race/ethnic group. Describe your findings. (*FAMILY9.DAT*)

Exercise 14 In 1950, the typical married couple family was supported by a single male earner. Is this true in 1990? What percentage of 1990 married couple families are headed by male single-earners? Dual earners? (*FERNTYP9.DAT*)

■ Create a pie chart for married couples with divisions for single-earner females, single-earner males, dual earners, and no-earners.

Earner Distribution of Married Couple Families

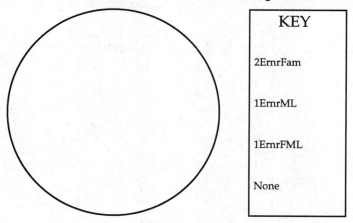

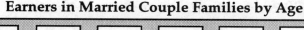

Exercise 15 How does the distribution of types of earners in married couples vary by age groups in 1990? What might account for these differences? (*FERNTYP9.DAT*)

■ Create a stacked bar chart with bars for each age group. For each age group, stack by the percentage of single-earner females, single-earner males, dual earners, and no-earner families in 1990.

Earners in Married Couple Families by Age

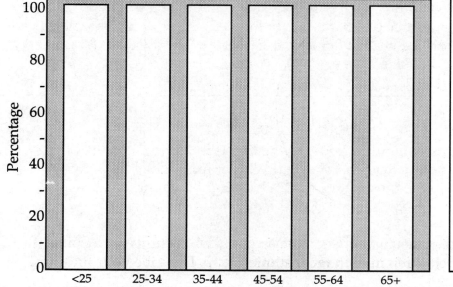

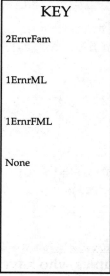

Exercise 16 According to the 1990 census, did the earning arrangement of married couples differ between race/ethnic groups? Which race/ethnic group had the highest percentage of dual earner married couples? The lowest? What are some possible differences between dual earner and single earner families? (*FERNTYP9.DAT*)

■ Create a stacked bar chart with bars for each race/ethnic group; stack by the percentages of dual earner, single-earner male, single-earner female, and no-earner families in 1990.

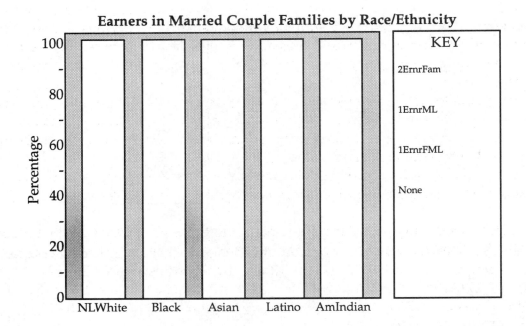

Exercise 17 Are married couple families less likely to be living in poverty than other family types? Using 1990 data, examine the percentage of households living in poverty among married couples, female-headed families, and male-headed families. (*FAMILY9.DAT*)

■ On your own, create three pie charts one for married couples, one for female-headed families, and one for male-headed families. In each pie, make divisions for poverty and non-poverty households.

𝒟iscussion Questions

1. In the 1950s, most families reflected the Ozzie and Harriet image of a married couple family headed by a single-earner male. How do current trends compare to the those of the 50s? What might account for these changes?

2. What is the relationship between household composition and income? Are married couples necessarily better off economically?

𝒟. Male and Female Headed Families

Most male or female headed family households are single parent families. The number of male-headed families has grown over the years, but they remain a small percentage of all family households. Most male-headed families are comprised of single fathers living with their children. Some of these fathers are widowers, but most are divorced.

While courts have become more sympathetic towards father-custody, most children with divorced parents live with their mothers. This practice has resulted in a surge of female-headed family households, most of which are comprised of single mothers living with their children. Many of these mothers are divorced and work to support their children.

Single parenthood has generated a lot of research and debate. Some argue that children living with single parents do not get all of the support they need. Others argue that living in a household filled with marital conflict is more detrimental than living in a single parent household.

In the following exercises you will look at the overall trends, race/ethnicity composition and economic well-being of male and female-headed families.

Exercise 18 Considering overall trends in household types since 1950 (refer to section A, exercise 3), discuss the trends in male-headed families since 1950. (*HH5090.DAT*)

Exercise 19 Focusing on 1990, determine which race/ethnic groups have the highest and lowest percentage of male-headed families. Discuss the differences between groups and what might account for the variation. (*FAMILY9.DAT*)

■ Create a bar chart with bars for each race/ethnic group indicating what percentage of family households is comprised of male-headed families.

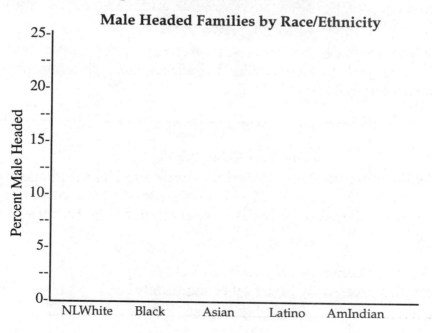

Male Headed Families by Race/Ethnicity

Exercise 20 Do male-headed families tend to be above the poverty level regardless of race/ethnicity? (*FAMILY9.DAT*)

■ Create a stacked bar chart with bars for each race/ethnic group; stack by male-headed families living above and below poverty in 1990. (Hint: Combine categories: NearPoor, Middle, Comf.)

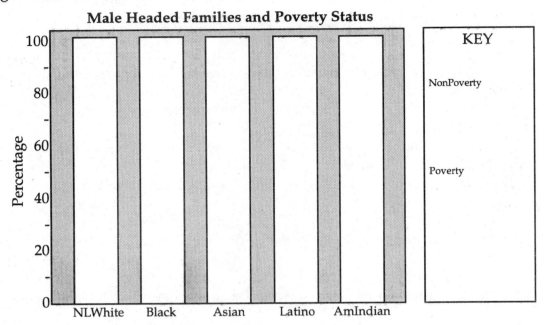

Male Headed Families and Poverty Status

Exercise 21 How has the number of female-headed families changed since 1950? What number of all family households were female-headed families in 1950? In 1990? What might account for the changes over time? (*HH5090.DAT*)

■ Create a line graph indicating the number of female-headed families for each year.

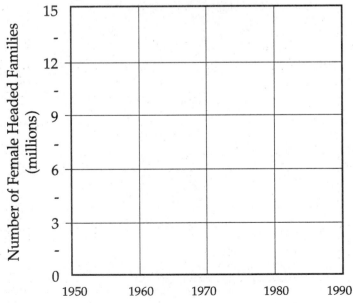

Female Headed Families, 1950 to 1990

■ (Part 2) Create a stacked bar chart with bars for each year; stack by female-headed and other family households (do not include nonfamilies in your calculations).

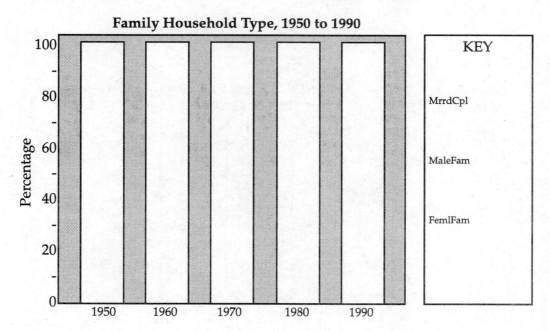

Exercise 22 What percentage of female-headed families have young children (under 6)? Do these percentages vary by race/ethnicity? (*FAMILY9.DAT*)

■ Create a bar chart with bars for each race/ethnic group. For each group, indicate the percentage of female-headed families with children under the age of 6 in 1990.

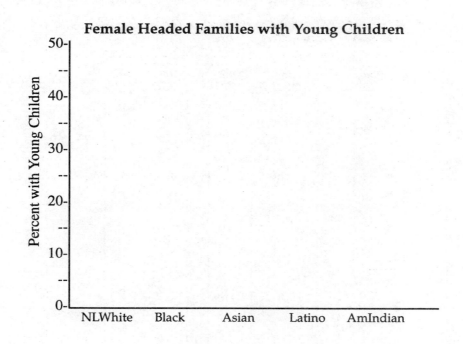

Exercise 23 How does the age distribution of female-headed families vary between race/ethnic groups in 1990? Describe any significant findings and what might account for the differences in age across race/ethnic groups. (*FAMILY9.DAT*)

■ Create a stacked bar chart with bars for each race/ethnic group; stack by age categories.

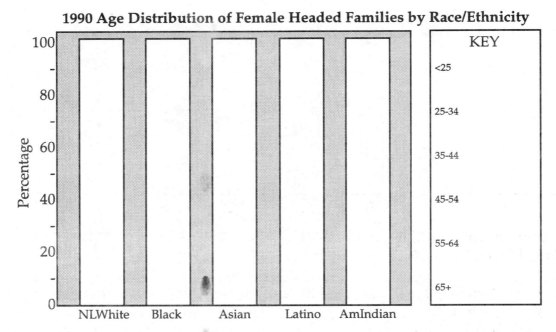

1990 Age Distribution of Female Headed Families by Race/Ethnicity

Exercise 24 What percentage of female-headed families were in poverty in 1990? Is race/ethnicity a factor in the poverty status of female-headed families? (*FAMILY9.DAT*)

■ Create a stacked bar chart with bars for each race/ethnic group; stack by the percentage of female-headed families living above and below poverty in 1990.

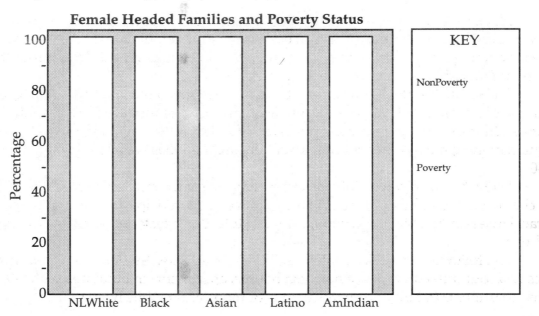

Female Headed Families and Poverty Status

Exercise 25 Write a brief summary comparing female-headed and male-headed families in the 1990s.

---

*D*iscussion Questions

1. Why do you think single parenthood has become more prevalent in the United States? Discuss the implications.

2. If 1990 trends in family composition and race/ethnic group differences persist, how might they affect the U.S. population twenty years from now? Do you think changes in family formation and marriage trends affect the well-being of children? Explain.

E. Housing and Households

Household and housing trends are interrelated. If household formation increases, the demand for housing increases. When a local population is mostly comprised of young, single adults, the type of required housing is very different from that of a population mostly comprised of large families.

The small, "starter" houses built for baby boomers in the late 1950s are a good example of housing that was specifically designed to accommodate demographic changes. In the face of skyrocketing marriage and birth rates, and an improved economic situation, the number of houses built during the 1950s was double that of the 1940s.

While household type influences housing construction, it's also true that housing characteristics influence who is likely to move in. Studio apartments don't tend to attract large families. Similarly, young, single adults don't seek out houses with five bedrooms.

The following exercises explore the relationship between household characteristics and four different types of housing: houses, apartments in buildings with 2-9 units, apartments in buildings with 10 units or more, and mobile homes.

Exercise 26 Which types of housing are most common for U.S. households in 1990? Does housing type differ by race/ethnicity? (*HOUSNG9.DAT*)

■ Create a pie chart showing the percentage of each housing type in 1990.

Housing Type Distribution in 1990

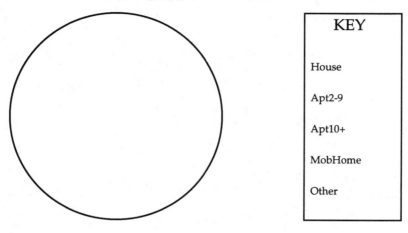

KEY

House

Apt2-9

Apt10+

MobHome

Other

■ On your own, create a stacked bar chart with bars for each race/ethnic group; stack by the percentage of housing types in 1990.

Exercise 27 Look at how 1990 housing types varies by the age of the householder. What might account for these differences? (*HOUSNG9.DAT*)

■ Create a stacked bar chart with bars for each age group; stack by housing types in 1990.

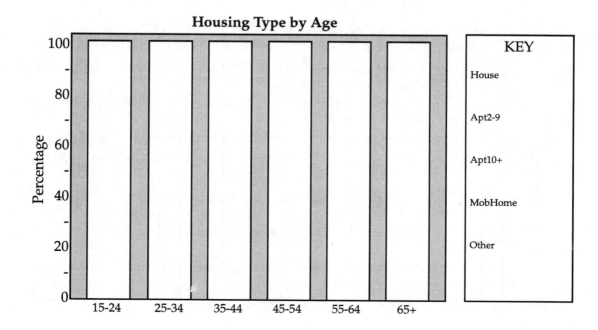

Exercise 28 Focusing on 1990, determine whether families and nonfamilies tend to live in the same kind of housing. Account for any differences. (*HOUSNG9.DAT*)

■ Create two pie charts, one for families and one for nonfamilies. In each chart, show the housing type distribution.

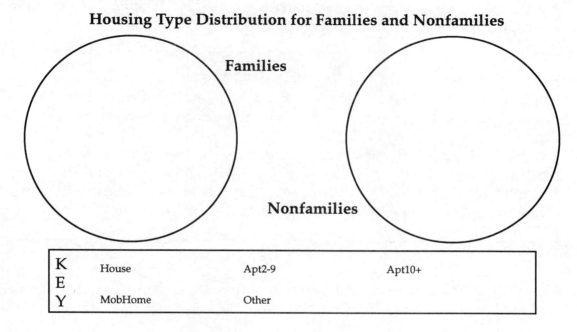

Housing Type Distribution for Families and Nonfamilies

Families

Nonfamilies

| K
E
Y | House | Apt2-9 | Apt10+ |
| | MobHome | Other | |

Exercise 29 Using 1990 data, determine whether the housing type of families varies by race/ethnicity. Describe your findings. (*HOUSNG9.DAT*)

■ Create a stacked bar chart with bars for each race/ethnic group; stack by the family housing type distribution. (Hint: Omit Nonfamily households.)

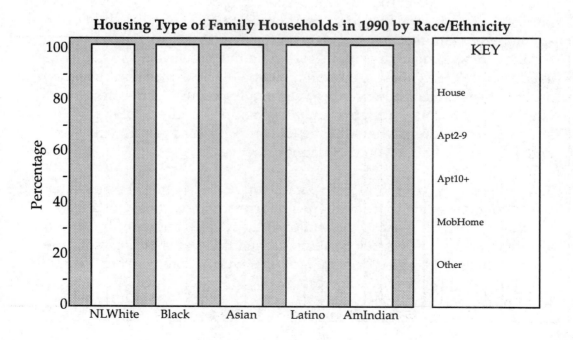

Housing Type of Family Households in 1990 by Race/Ethnicity

Exercise 30 What is the relationship between home ownership and family composition? Are female-headed families as likely to own a home as married couples? (*HOUSNG9.DAT*)

■ On your own, create three pie charts, one for female-headed families, one for male-headed families, and one for married couples. In each pie, show the percentage of homeowners and non-homeowners.

---

*D*iscussion Questions

1. Home ownership is considered a keystone in the American dream. Do you think this aspiration is justified? Do you think affordable housing in a right?

2. How does age influence housing and living arrangements? Consider a "typical" person at each of the following ages: 10, 20, 30, 40, 50, 60, and 70+. How might his or her housing arrangements change over time? What factors might cause people or households to change their type of residence?

THINK*tank*

1. Single-parent families are generally less economically well off than married couple families. However, does the age and education level of the household head affect the economic well-being of female and male headed families? Provide evidence to support or refute the claim that children in single-parent families may be better off than children in married couple families, depending on a parent's age and education level. Are there other aspects of *well being* (beside economic) that would be important to explore in order to determine differences between single-parent and married couple families? If you could design a study to examine the effects of family structure on children, what other aspects of well-being would you want to know about, and how would you measure them?

2. While reading your local newspaper, you notice a cantankerous letter to the editor suggesting that certain neighborhoods should be zoned for families only. According to the author of the letter, all nonfamily households are composed of young college students, and their late hours disrupt the peace of neighborhoods. Do you think this type of zoning is fair? Why or why not? Write an informative response to the letter which illustrates who would be affected by such zoning.

Many Americans don't have an accurate understanding of the current state of poverty in the United States. While overall poverty levels have decreased significantly since the late 1940s, they have recently been on the rise. Contrary to myths about who is living in poverty and why, the picture of poverty in America illustrates that it is an important and universal issue.

It is necessary to look at how poverty is defined before attempting an analysis of the situation. The federal government bases its definition of poverty upon a Department of Agriculture nutritional study conducted in 1961. This study established that the average non-farm family spent one-third of its income on food. In light of this information, the researchers developed an "Economy Food Plan" which established the minimum cost for fulfilling basic nutritional needs. The poverty level was then calculated by multiplying the cost of the "Economy Food Plan" by three. Families with earnings below three times the amount of the "Economy Food Plan" are considered to be living in poverty. Naturally, the calculations take inflation and family size into account.

While the results of this study are still used to measure poverty today, many politicians and researchers feel this measurement of poverty is flawed. Critics of the measurement argue that it is outdated, does not consider changes in the nation's

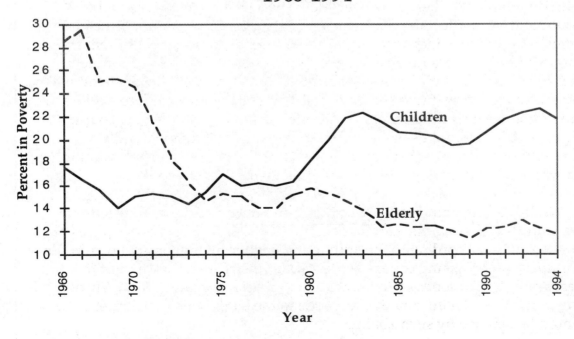

Poverty Rate for Children and the Older Population 1966-1994

Source: Population Reference Bureau

KEY concepts

Poverty Status of a Family At its core, the Federal Government's poverty classification is a family-based measure. A family's poverty status ("in poverty" versus "not in poverty") is determined on the basis of the family's total income and whether or not it falls below the "poverty cutoff" (the income threshold below which is considered poverty income). The "poverty cutoff" is based on the assumed costs of a family's nutritional needs and differs according to a family's size and composition. For example, the 1990 Census used a "poverty cutoff" of $12,675 for a family of four. This cutoff is lower for a family of three, and higher for a family of five. Because poverty status is based on the combined income of the family, considerations such as the number of earners in a family are important in affecting its poverty status.

Poverty Status of a Person Although poverty status is a family-based concept, persons can also be classified as "in poverty" or "not in poverty" whether they live in family or nonfamily households. Persons living in family households are simply assigned the poverty status of their family. However, persons residing in nonfamily households are treated as individuals. Their poverty status is determined on the basis of individual level "poverty cutoffs". When using poverty status information for persons, bear in mind that, for persons who live in families, the statistics reflect their family's incomes rather than their individual income.

Income Relative to Poverty Cutoff While the "poverty cutoff" distinguishes between poverty and nonpoverty incomes, it can also be used to interpret different levels of nonpoverty income. This involves expressing actual income as a ratio to the "poverty cutoff". For example, if the poverty cutoff for a family of four is $12,675, a family income up to 1.5x the poverty cutoff, ($19,012) they might be considered "near-poor", and a family more than 5x the poverty cutoff ($63,375) might be considered "comfortable". For a smaller family with a lower poverty cutoff, a "comfortable" income will also be lower; for a larger family it will be somewhat higher. Adopting a formula by Judith Treas and Ramon Torrecilia (see *References for Topics*), we use the following "Income to Poverty Cutoff" Conversion:

> *poor* income below the "poverty cutoff"
> *near-poor* income between 1x and 1.5x the "poverty cutoff"
> *middle* income between 1.5x and 5x the "poverty cutoff"
> *comfortable* income above 5x the "poverty cutoff"

OTHER concepts

Family Household (Topic seven)   **Region** (Topic three)
Race/Ethnicity (Topic two)   **City-Suburb-Nonmetropolitan**
Education (Topic two)   (Topic one)

consumption patterns, and does not account for regional differences in the cost of living. For example, food is much more expensive in urban areas than rural areas.

Regardless of how poverty is measured, the fact remains that it is still a large problem. In the face of this problem, we will look at who is in poverty, why they are in poverty, and what is being done to assist them.

Many people have a vague concept of "welfare" as the government's solution to poverty. In the United States, "welfare" is actually comprised of several different federal programs, each with distinct qualifying guidelines. Many of these programs were developed during the New Deal Era, including Social Security Retirement, Social Security Disability Insurance, Unemployment Insurance and Aid to Families with Dependent Children (AFDC). In the 1960s, President Johnson's "War on Poverty" gave birth to a second wave of reforms: Supplemental Security Income, the School Lunch Program, Medicaid, Medicare and the Food Stamp Program.

Despite government assistance programs, societal and economic changes in recent decades have led to an increase in the number of people in poverty. During the late 70s and early 80s, real wages were depressed due to a recession and an influx of baby boomers looking for work. Furthermore, in the face of tougher competition, those with lower education levels had difficulty securing employment. In addition to economic trends, the changing composition of American family households has influenced poverty levels. Over the past three decades, an increase in divorce rates has led to an increase in single parent families, the vast majority of which are female-headed. The increase in single mothers, coupled with the lower overall earnings of females, has created a unique trend known as the *feminization of poverty*.

In the following exercises, you will explore overall poverty trends, and focus on poverty patterns in terms of race/ethnicity, family type, gender, age, educational attainment and geographic location. As you work through these exercises, consider the economic trends and historical events which may have influenced the current state of poverty in America.

$\mathcal{A}$. Poverty and Race/Ethnicity

Despite the Civil Rights movement and changing attitudes, there is still a strong relationship between a household's race/ethnicity and its likelihood of being classified below the poverty level. In the following exercises, you will look at overall poverty trends in terms of race/ethnicity, and focus on the state of poverty among different race/ethnic groups in 1990. As you complete these exercises, recall other factors connected to race/ethnicity, such as educational attainment and earnings, which may partially account for your findings.

Exercise 1 Look at the percentage of black and nonblack family households in poverty from 1970 to 1990. Describe your findings. (*FPOV7090.DAT*)

■ Create a line graph with a line for blacks and a line for nonblacks; for each year, indicate the percentage of family households in poverty.

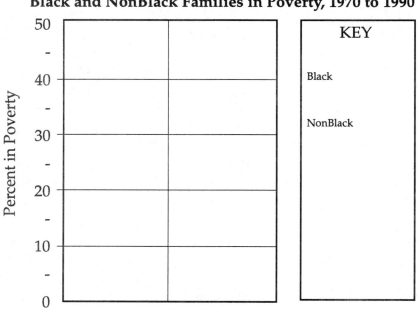

Black and NonBlack Families in Poverty, 1970 to 1990

KEY

Black

NonBlack

Exercise 2 Focusing on 1990, consider the race/ethnicity distribution of poverty and non-poverty families. Which race/ethnic group comprises the majority of families in poverty? Non-poverty? Describe any significant differences. (*FAMILY9.DAT*)

■ Create two pie charts, one for poverty and one for non-poverty, with divisions for each race/ethnic group. (Hint: Combine NearPoor, Middle, and Comf categories.)

Race/Ethnicity Distributions of Families

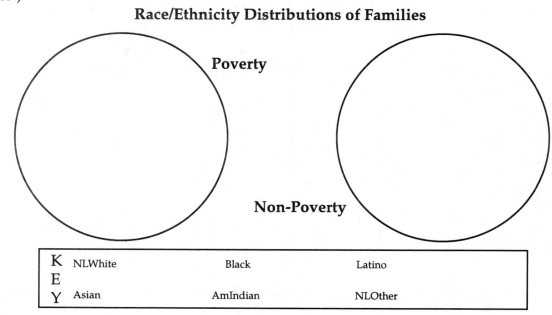

Poverty

Non-Poverty

K E Y	NLWhite	Black	Latino
	Asian	AmIndian	NLOther

Exercise 3 Consider the previous exercise from a different angle. Using 1990 data, look at the percentage of families in poverty in each race/ethnic group. Describe the differences between race/ethnic groups. (*FAMILY9.DAT*)

■ Create a bar chart with bars for each race/ethnic group; for each group, indicate the percentage of families in poverty in 1990.

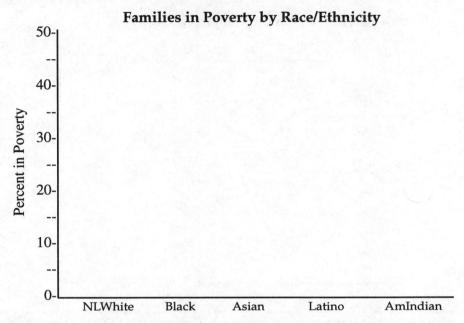

Families in Poverty by Race/Ethnicity

Exercise 4 Using 1990 data, look at the percentage of families in poverty within each specific Latino group. Which groups show the highest percentage of families in poverty? Lowest? What are some possible explanations for these differences? (*FPOVLAT9.DAT*)

■ Create a bar chart with bars for each specific Latino group; for each group, indicate the percentage of families in poverty in 1990.

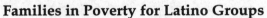

Families in Poverty for Latino Groups

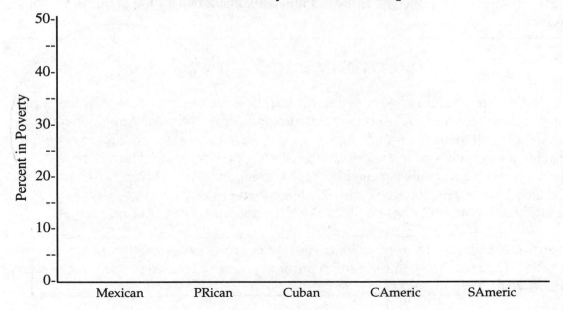

Exercise 5 Using 1990 data, look at the percentage of families in poverty within each specific Asian group. Which groups show the highest family poverty? Lowest? What are some possible explanations for these differences? (*FPOVASN9.DAT*)

■ Create a bar chart with bars for each specific Asian group; for each group, indicate the percentage of families in poverty in 1990.

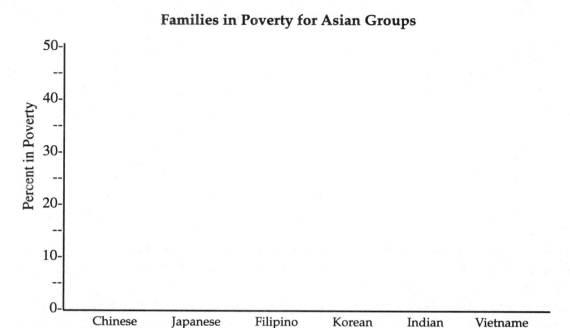

Families in Poverty for Asian Groups

$\mathcal{D}$iscussion **Questions**

1. Can you think of any historical occurrences that may have affected trends in poverty?

2. Did the results from exercise 2 surprise you? How do the results differ from contemporary myths about poverty? Can you think of explanations for the patterns shown in exercise 2?

$\mathcal{B}$. Poverty and Family Type

Over the last fifty years, economic trends as well as marriage, divorce and childbearing rates have altered the structure and composition of American families. While you will focus on female-headed families in the next section, these exercises will provide an overview of the relationship between poverty and all family types.

Are certain families more likely to be living in poverty? Does this vary by race/ethnicity? As you explore the connection between poverty and family type, recall other factors related to family type which may also influence socioeconomic status.

Exercise 6 Using data from 1970 to 1990, look at poverty in different types of families. Describe the trends. Which family types experience the highest poverty? The lowest? (*FPOV7090.DAT*)

■ Create a bar chart with side by side bars for each family type; for each year, indicate the percentage of each family type living in poverty.

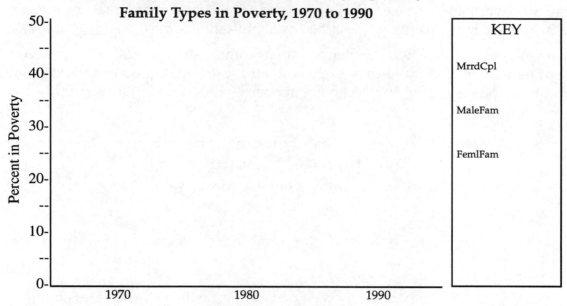

Family Types in Poverty, 1970 to 1990

Exercise 7 Using data from 1970 to 1990, look at poverty in black and nonblack families. Describe the trends. Does poverty seem more influenced by race/ethnicity or family type? (*FPOV7090.DAT*)

■ Create two line graphs, one for blacks and one for nonblacks. In each graph, draw three lines, one for each family type. Indicate the percentage of each family type in poverty for each year.

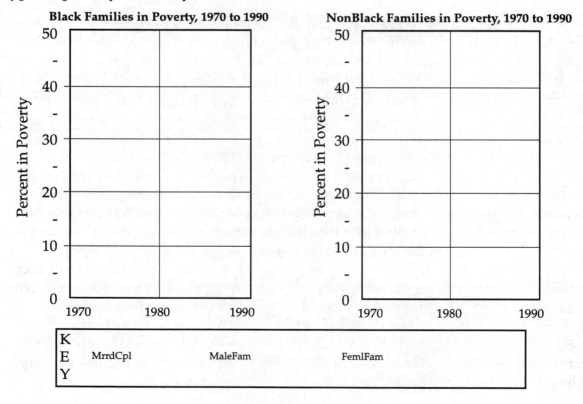

Black Families in Poverty, 1970 to 1990

NonBlack Families in Poverty, 1970 to 1990

Exercise 8 Do you think race/ethnic differences in poverty are related to race/ethnic differences in family structure? Focusing on 1990, consider the percentage of families living in poverty in each race/ethnic group and the percentage of female-headed families in each race/ethnic group. Describe any significant findings. (*FAMILY9.DAT*)

■ On your own, create two bar charts. The first chart should indicate the percentage of female-headed families in each race/ethnic group. The second chart should indicate the percentage of families living in poverty in each race/ethnic group.

---

*D*iscussion **Questions**

1. Why do you think different family types experience varying levels of poverty?

2. To what extent is family type a factor in the increased rates of poverty among blacks?

3. What factors other than family type may influence the poverty rates among minority groups?

C. Gender and the Feminization of Poverty

Over the last thirty years, researchers have been observing the relationship between gender and poverty. This relationship has become known as the *feminization of poverty*. In order to understand the emergence of this trend, you must first consider the relationship between household composition and changing social circumstances. Half of female-headed families living in poverty have undergone a household composition change, usually resulting from divorce. Thus, as divorce rates have been increasing over the past thirty years, so has the number of female-headed family households. Many of these female-headed family households are completely responsible for supporting children. This responsibility has become more challenging as the cost of child care and health care increases. In addition to high divorce rates, there has been an increase in out-of-wedlock births which, like divorce, frequently leave the mother as the only means of support.

Exercise 9 What accounts for the fact that female-headed families constitute the majority of all families in poverty? Using data from 1970 to 1990, look at what percentage of all family households are female-headed. Then, look at the percentage of all family households in poverty which are female-headed. Describe your findings. (*FPOV7090.DAT*)

■ Create two bar charts, one for all female-headed families and one for female-headed families in poverty. In the first chart, indicate what percentage of all family households are female-headed. In the second chart, indicate the percentage of all family households in poverty which are female-headed.

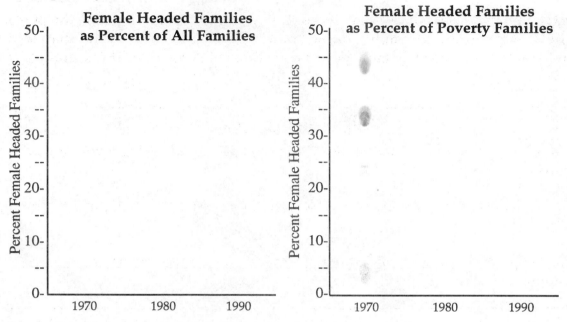

Exercise 10 On your own, repeat the previous exercise for black female-headed families only. Compare your findings to those for the overall population. (*FPOV7090.DAT*)

Exercise 11 How does the poverty of female-headed families vary by race/ethnicity? Compare the overall percentage of female-headed families to the percentage of female-headed families in poverty for each race/ethnic group in 1990. Describe any differences between groups. (*FAMILY9.DAT*)

■ On your own, create two bar charts, one for female-headed families and one for female-headed families in poverty. In the first chart, indicate the percentage female-headed families comprise of all family households for each race/ethnic group. In the second chart, indicate the percentage female-headed families comprise of all family households in poverty for each race/ethic group.

Exercise 12 For this exercise, we are going to focus on persons rather than families (See "Key Concepts" for the distinction between the poverty status of families and persons). Looking only at persons 18 years old and older in 1990, consider how poverty differs between gender and race/ethnic groups. Which groups have the highest percent poverty? The lowest? Offer explanations for your findings. (*PPOVGEO9.DAT*)

■ Create a bar chart with side by side bars for men and women. For each race/ethnic group, indicate the percentage of men and women living in poverty.

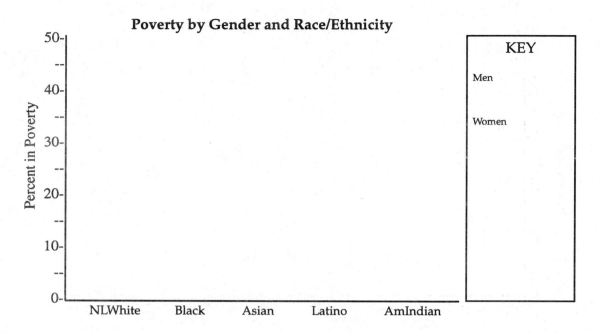

Poverty by Gender and Race/Ethnicity

$\mathcal{D}$iscussion Questions

1. Do you agree that "feminization of poverty" is an accurate label for recent trends in poverty? Why or why not?

2. How does the feminization of poverty affect race/ethnic groups differently?

$\mathcal{D}$. Poverty Among Children and the Elderly

Over the last forty years, social assistance programs, such as Social Security, have helped many elderly people stay above the poverty level. The story for children, however, has been one of reduced benefits and small overall reductions in poverty levels. Poverty trends among children can be partially accounted for by the increase in female-headed families, which you explored earlier.

In this section and the next we will focus on the poverty situation of *persons*, rather than families (see "Key Concepts"). You will look at the poverty of persons in all age groups, but will focus on the elderly and children. As you examine the relationship between age and poverty, recall other factors related to age, such as education, marital status and earnings.

Exercise 13 Focusing on 1990, compare the percentage of people in poverty in each age group. Which ages have the highest percentage of people in poverty? Lowest? What might account for your findings? (*PPOVGEO9.DAT*)

■ Create a bar chart indicating the percentage of people in poverty in each age group in 1990.

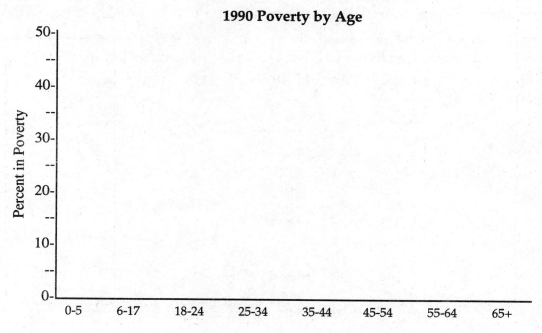

1990 Poverty by Age

Exercise 14 Using 1990 data, look at the percentage of children (ages 0-17) and elderly (ages 65+) in poverty within each race/ethnic group. Describe your findings. (*PPOVGEO9.DAT*)

■ Create a bar chart with side by side bars for children and the elderly. For each race/ethnic group, indicate the percentage of children and the elderly in poverty in 1990. (Hint: Combine age categories 0-5 with 6-17 for children.)

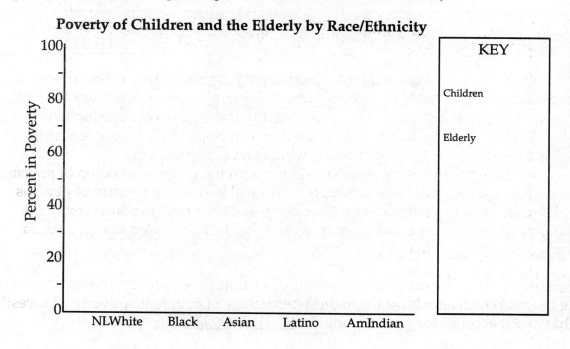

Poverty of Children and the Elderly by Race/Ethnicity

Exercise 15 Focusing on 1990, compare the percentage of elderly men living in poverty to the percentage of elderly women living in poverty. Is this gender gap significantly different from the overall gender gap in poverty? (*PPOVEDU9.DAT*)

■ Create two pie charts, one for elderly men and one for elderly women. In each pie, show the percentage living in poverty and the percentage not in poverty.

Poverty Status of Elderly Men and Women

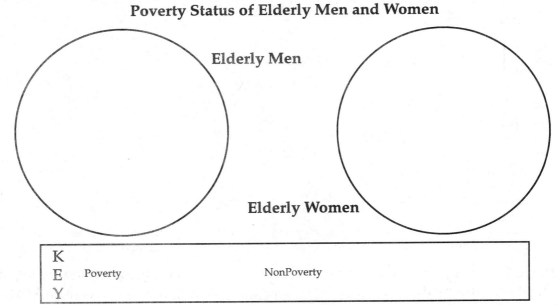

Elderly Men

Elderly Women

| K E Y | Poverty | NonPoverty |

---

*D*iscussion **Questions**

1. Why do children and the elderly have different levels of poverty? Compare these two age groups to each other and the overall population.

2. Do you think that future cohorts will follow the same age-related pattern of poverty?

E. Education and Poverty

Over the past few decades, the American labor market has undergone significant restructuring. Blue collar positions in industry, once the mainstay of our economy, have decreased significantly. Technological changes, coupled with a downsizing of industry, have led to a demand for workers with higher education levels. In the face of tougher competition, those with lower education levels have had difficulty securing positions.

This section also focuses on persons rather than families. You will examine the relationship between an individual's poverty and educational attainment. As you examine this relationship, recall other factors related to education, such as race/ethnicity and gender.

Exercise 16 Focusing on 1990, look at the relationship between educational attainment and poverty for people ages 25-34. Describe any significant findings. (*PPOVEDU9.DAT*)

■ Create a bar chart for 25-34 year olds indicating the percentage in poverty for each education level in 1990.

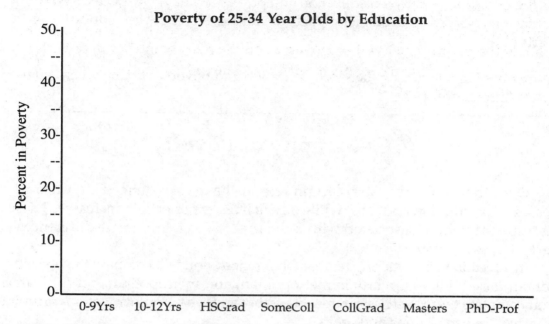

Poverty of 25-34 Year Olds by Education

Exercise 17 Using 1990 data, examine the relationship of education and poverty in terms of race/ethnicity. Looking only at 25-34 year old college graduates and those with less than a high school education, determine the percentage in poverty in each race/ethnic group. Describe any significant differences. *(PPOVEDU9.DAT)*

■ Create a bar chart with side by side bars, one for 25-34 year old college graduates and one for those with less than a high school education. Indicate the percentage of people in poverty for each education level in each race/ethnic group.

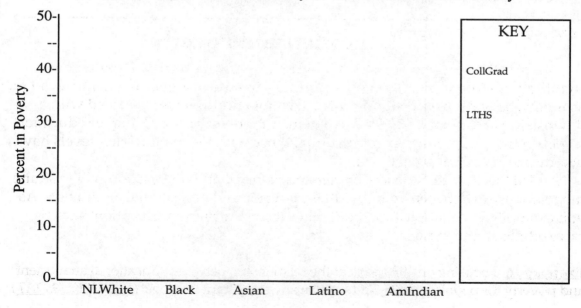

Poverty Among 25-34 Year Olds by Education and Race/Ethnicity

KEY

CollGrad

LTHS

F. Geography of Poverty

Over the past fifty years, certain regions have experienced significant increases in population and jobs, while others have experienced decreases. These shifts, especially those associated with industrial downsizing, create differences in poverty across regions.

In addition to considering regional differences, you will compare the percentage of families living in poverty in metropolitan, non-metropolitan and rural areas. As you complete these exercises, think about how and why employment opportunities may vary between regions and areas.

<u>Exercise 18</u> Focusing on 1990, show how the percentage of poverty families differs across city, suburban, and non-metropolitan areas. Which area has the highest percentage of families in poverty? The lowest? Did you expect to find a different pattern? *(FPOVGEO9.DAT)*

■ Create a bar chart indicating the percentage of family households in poverty in each area.

Percent in Poverty by Geography

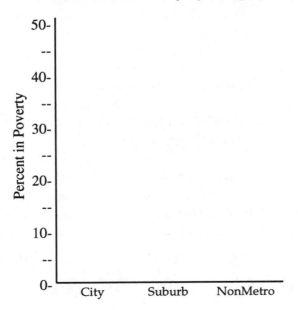

Exercise 19 Is the percentage of family households in poverty in cities, suburbia, and non-metropolitan areas similar for blacks and nonblacks? Focusing on 1990, compare the percentage of black and nonblack families in poverty in city, suburban, and non-metropolitan areas. *(FPOVGEO9.DAT)*

■ Create two bar charts. In the first chart, indicate the percentage of black families in poverty for each area. In the second chart, indicate the percentage of nonblack families in poverty in each area.

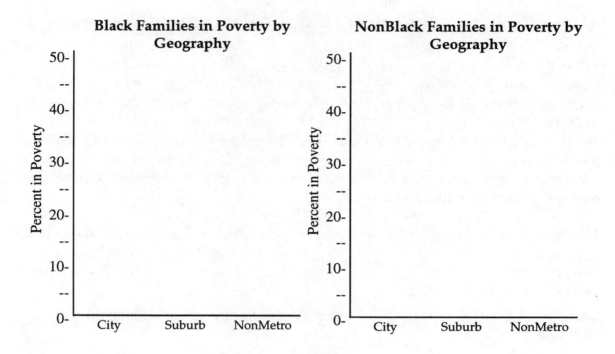

Exercise 20 On your own, select another race/ethnic group you would like to focus on and repeat the previous exercise for that particular group. *(FPOVGEO9.DAT)*

---

*D*iscussion Questions

1. Are the conditions associated with poverty in cities similar to the conditions associated with poverty in non-metropolitan areas?

2. How do you think the percentage of family households in poverty varies across regions in the United States? Which region, Northeast, South, Midwest, or West, do you think has the highest percentage of family households in poverty?

THINK tank

1. Examine the relationship between poverty and age, gender, race/ethnicity, household structure, education, and metropolitan geography. Discuss strategies you think would help to reduce poverty. Make two or three recommendations for policy changes at the local, state, and federal levels. What are some of the benefits and shortcomings to each of these strategies?

2. What has happened to the level of poverty of the old the young over the last few decades? Why have these changes occurred? Given that governments allocate limited resources to combat poverty, is it better to allocate funds to impoverished children, or to impoverished elderly? If you were a political candidate running for national office, what social, economic, demographic, and political factors would you consider before taking your position on this policy dilemma? Use data to help explain your consideration and justify your policy position.

CHILDREN
topic nine

Teachers, politicians and even musicians are fond of reminding us that "children are our future." While this statement has become a cliché, we can, to some extent, predict what America may be like in the future by looking at the children of today. How have changes in household structure, gender roles and employment opportunities affected children's lives? What does the future of America look like?

Over the last forty years, attention to children's issues in public policy waxed and waned in relation to the relative size of the child population. During the late 1950s and early 1960s, public officials built new schools, hired more teachers, and provided more funding to educate the swelling ranks of the baby boomers. In addition to education and housing policies designed to aid middle class families with kids, other policies were directed at children living in poverty. Title I of the Elementary and Secondary Education Act, part of the Johnson Administration's "War on

Poverty", committed substantial amounts of federal aid to poor, underachieving children. Similarly, Head Start, Aid to Families with Dependent Children (AFDC), and Women and Infant Children (WIC) were designed to help address child poverty. While these programs still exist, somewhat fewer programs for children have been developed during the past decade. This shift in program development is partially due to the decrease in the children's cohort size. In other words, public policy focused more on the needs and conditions of children when they represented a larger percentage of the United States population. Interestingly, as the size of the elderly increases, issues like social security and health care are in the policy spotlight.

In the following sections, you will explore the current state of America's children. Who are they? How do they live? With whom do they live? Are some living in better conditions than others? Who are the advantaged and disadvantaged children, and why do these disparities exist? The exercises begin with a discussion of childhood trends since 1950, and then provides a snapshot of the race/ethnicity distribution and immigration status of children in 1990. Subsequent exercises examine critical changes in family structure, and children living in poverty. You will also take a brief look at children and schooling in the 1990s. As you work through these exercises, pay particular attention to the differences between children, and consider the effects that these differences may have on their future, and your own.

KEY concepts

Child Population Persons 0-17 years old.

School Aged Population Persons 6-17 years old.

Public/Private Schools Persons attending school indicated that they were attending public school or private school:

   Public school any school or college controlled or supported by a local, county, state, or federal government.

   Private school schools supported and controlled primarily by religious organizations or other private groups.

OTHER concepts

Race/Ethnicity (Topic two)   Family Households (Topic seven)
Immigration Status (Topic three)   Education (Topic two)
Poverty Status (Topic eight)   English Language Proficiency (Topic
Income Relative to Poverty Cutoff three)
(Topic eight)

A. The Child Population

In addition to looking at the overall size of the child population, it is important to examine children as a percentage of the total U.S. population. As this percentage increases, the child dependency ratio — the number of children who need care, compared to the number of adults able to provide care — tends to increase as well. In practical terms, this often means that more kids have to share fewer resources. However, in the arena of politics, a large cohort can sometimes wield more influence.

The line graph below shows the number of children (ages 0-17) living in the U.S. between 1950 and 1990. Notice how the size of the current child population compares with previous decades. What might account for the overall decline in the child population? Do you think attitudes about children have changed since 1950? What are the social and economic implications of having children as a large or small percentage of the total population? Do you think it is more advantageous to be born into a large cohort or a small one? Why?

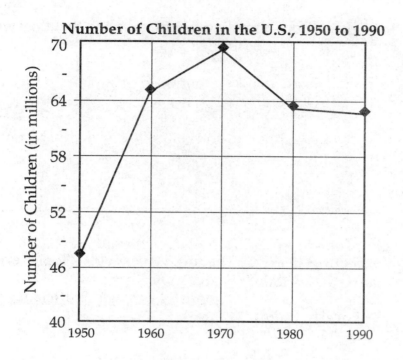

Number of Children in the U.S., 1950 to 1990

$\mathcal{B}$. Race, Ethnicity, Immigration and Children

One of most significant changes in the child population over the last few decades has been a dramatic change in its race/ethnicity distribution. Although the U.S. has long been home to a diverse group of people, that diversity is increasing. High birth rates among some race/ethnic groups account for a great deal of the shift, but another factor has been the significant number of Latin Americans and Asians who have immigrated to the United States since 1970. As you look at the race/ethnicity distribution of the child population, it is important to note that some states have more diverse populations than others.

Exercise 1 Using 1990 data, examine the race/ethnicity distribution of the child population (ages 0-17) and people ages 45-54. What differences do you see? Why is it useful to look at these age groups? (*PPOVGEO9.DAT*)

■ Create a stacked bar chart with two bars, one for 0-17 year olds and one for 45-54 year olds; stack by race/ethnicity. (Hint: Combine ages 0-5 with 6-17.)

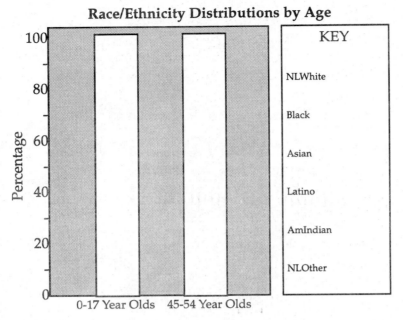

Race/Ethnicity Distributions by Age

KEY

NLWhite

Black

Asian

Latino

AmIndian

NLOther

Exercise 2 Do some race/ethnic groups have a larger percentage of their total population made up of children than other race/ethnic groups? Compare these percentages for each race/ethnic group. (*PPOVGEO9.DAT*)

■ Create a bar chart with bars for each race/ethnic group; for each group, indicate the percentage of children (persons ages 0-17).

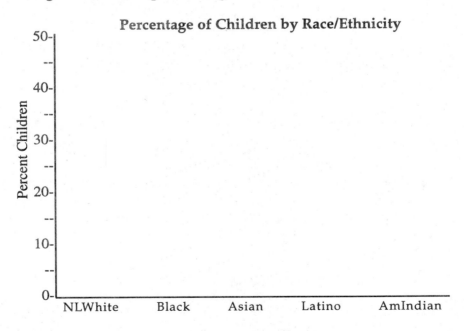

Percentage of Children by Race/Ethnicity

Exercise 3 On your own, repeat the previous exercise for each specific Latino and Asian group. (*ENGASN9.DAT, ENGLAT9.DAT*)

Exercise 4 Does family size vary between race/ethnic groups? Look at the family size distribution in each race/ethnic group. Describe any significant findings. (*CHLDPOV9.DAT*)

■ Create a stacked bar chart with bars for each race/ethnic group; stack by family size.

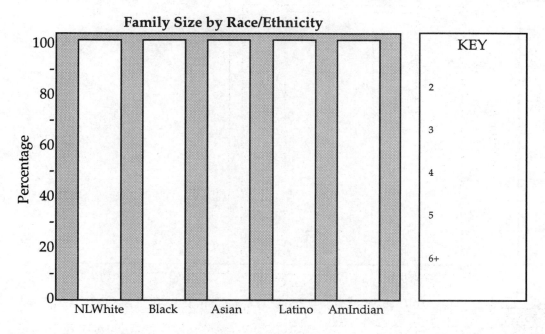

Exercise 5 According to population projections based on the 1990 census, what percentage of children (0-17) will be Asian, black, white, Latino, or American Indian race/ethnicity in 2010? (*POPPROJ9.DAT*)

■ Create a pie chart with divisions for your projected race/ethnicity distribution.

Projected Race/Ethnicity Distribution of the United States for 2010

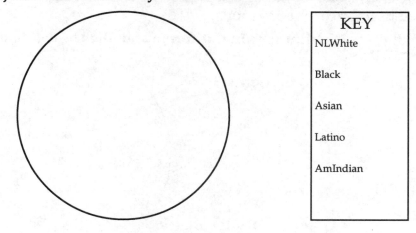

Exercise 6 What percentage of children living in the United States is foreign born? What percentage of children in each race/ethnic group is foreign born? Describe your findings. (*CHLDPOV9.DAT*)

■ Create a bar chart with a bar for the total child population and bars for each race/ethnic group; for each group, indicate the percentage of foreign born children.

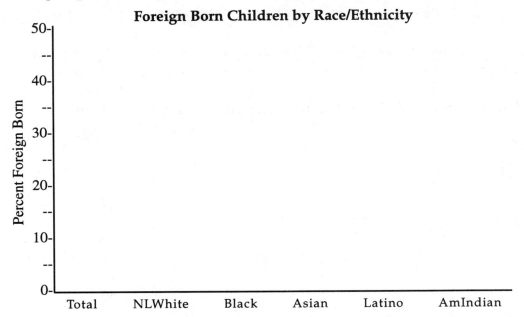

Foreign Born Children by Race/Ethnicity

Exercise 7 On your own, repeat the previous exercise for each specific Latino and Asian group. (*ENGLAT9.DAT, ENGASN9.DAT*)

Exercise 8 It is important to remember that exercise 6 refers to the U.S. as a whole. As you can probably guess, the picture of children in some states, for example California, does not reflect the overall national average. Do some race/ethnic groups account for a larger percentage of Californian children than others? Compare the race/ethnicity distribution of California children with that of the entire nation. (*POPPROJ9.DAT*)

■ Create two pie charts, one for the United States and one for California. In each pie, make divisions for race/ethnic groups.

Race/Ethnicity Distributions in California and the United States

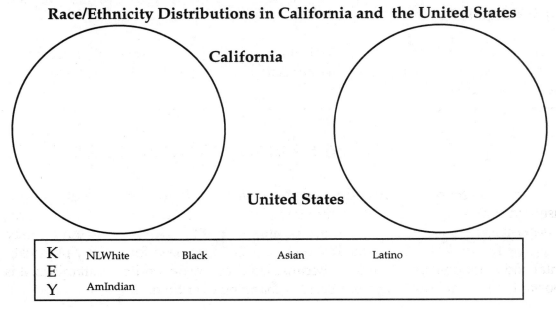

Exercise 9 Focusing on children living in California, examine the percentage of children in each specific Latino and Asian group. Which specific groups account for the greatest percentage of Asian and Latino children living in California? (*ENGLAT9.DAT*, *ENGASN9.DAT*)

■ Create two pie charts, one for Asian children living in California and one for Latino children living in California. In each pie, make divisions for the specific groups.

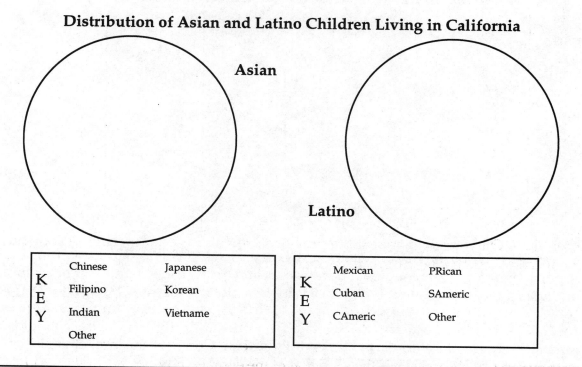

Distribution of Asian and Latino Children Living in California

Asian

Latino

K E Y	Chinese	Japanese
	Filipino	Korean
	Indian	Vietname
	Other	

K E Y	Mexican	PRican
	Cuban	SAmeric
	CAmeric	Other

---

*D*iscussion **Questions**

1. What factors have influenced the race/ethnicity distribution changes over the past few decades?

2. Why do some states have child populations that are more racially/ethnically diverse than others? Why aren't race/ethnic populations distributed evenly across the country?

C. Children's Family Structure

Previous chapters have addressed how changes in family structure affect household composition, socioeconomic status and labor force participation. However, this demographic shift is particularly important for children. Families are the primary providers of nurture, aid and other resources that children need for healthy physical, mental and emotional development. Because children are dependent on families, it is important to look at how family structure is changing over time.

Exercise 10 Keeping in mind the family type distribution we discussed in the Households and Families chapter, look at the percentage of children living in married couple families, female-headed families, and male-headed families in 1990. Describe your findings. (*CHLDPOV9.DAT*)

■ Create a pie chart for all children with divisions by each family type.

Family Type Distribution for Children in the U.S.

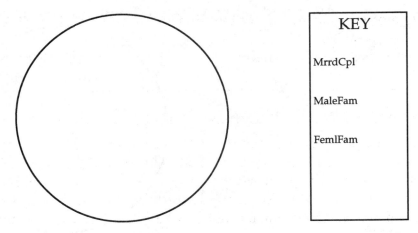

KEY

MrrdCpl

MaleFam

FemlFam

Exercise 11 Using 1990 data, determine the percentage of children in each race/ethnic group living in married couple, female-headed, and male-headed families. For example, what percentage of white children are in single-mother families? (*CHLDPOV9.DAT*)

■ Create a stacked bar chart with bars for each race/ethnic group; stack by family types.

Family Types by Race/Ethnicity

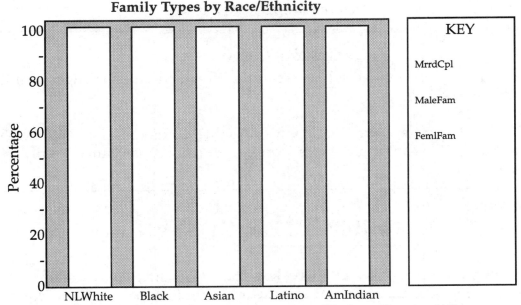

KEY

MrrdCpl

MaleFam

FemlFam

Exercise 12 As children get older, are they more likely to live in different types of families? (*CHLDPOV9.DAT*)

■ On your own, create three pie charts, one for each family type. In each pie, make divisions for young children (0-5) and older children (6-17).

Discussion Questions

1. Based on your knowledge from other chapters, do you think there has been an increase in the number of children living in single parent families over the past few decades? Why?

2. Do you think divorce is detrimental to children? Why/why not?

3. Why is single motherhood such a contentious issue in American politics?

4. Describe the trends and implications of more children living in single mother families.

D. Economic Well-Being of Children

There is little question that child poverty is a problem in America today, particularly for kids in some cities and rural areas. Compared to the elderly, children are much more likely to be poor. Moreover, poor children are very likely to become poor adults. Although it is clear that poverty affects a child's health, career aspirations and academic achievement, it is not clear how to resolve this seemingly intractable problem.

Exercise 13 What percentage of U.S. children live in poverty? How has that percentage changed since 1970? *(PPOV7090.DAT)*

■ Create a line graph showing the percentage of children in poverty for each year.

Children in Poverty 1970 to 1990

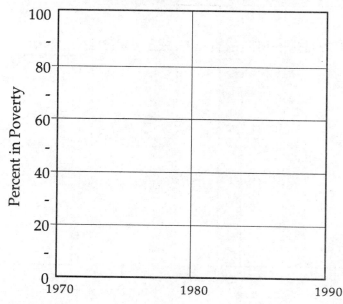

Exercise 14 What percentage of children in 1990 lived in families with comfortable incomes? (Note: See "Key Concepts" in the Poverty Chapter for a definition of comfortable incomes.) *(PPOVGEO9.DAT)*

■ Create a bar chart with two bars, one for children (ages 0-17) and one for the overall population (all ages). In each bar, indicate the percentage with comfortable incomes.

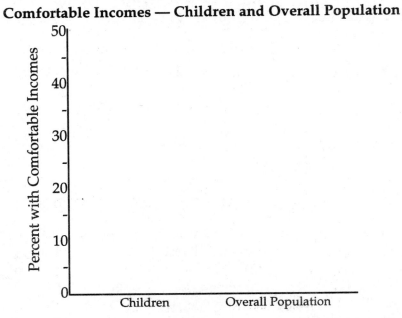

Comfortable Incomes — Children and Overall Population

Exercise 15 Using 1990 data, determine the percentage of children within each race/ethnic group who live in poverty or in families with comfortable incomes. For example, how many black children live in families with comfortable incomes? (*CHLDPOV9.DAT*)

■ Create a bar chart with side by side bars for poverty and comfortable incomes. For each race/ethnic group, indicate the percentage of children who are poor or comfortable.

Children in Poor and Comfortable Income Families by Race/Ethnicity

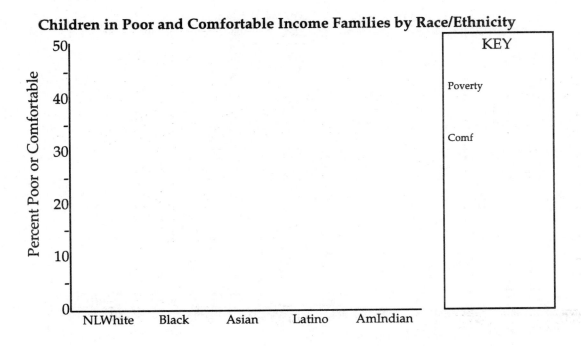

Exercise 16 According to the 1990 census, is there a substantial difference between the percentage of young children (0-5) in poverty and the percentage of older children (6-17) in poverty? Does the percentage of young and older children in poverty vary by race/ethnicity? (*CHLDPOV9.DAT*)

■ Create two pie charts, one for young children and one for older children. In each pie, make divisions for poverty and non-poverty.

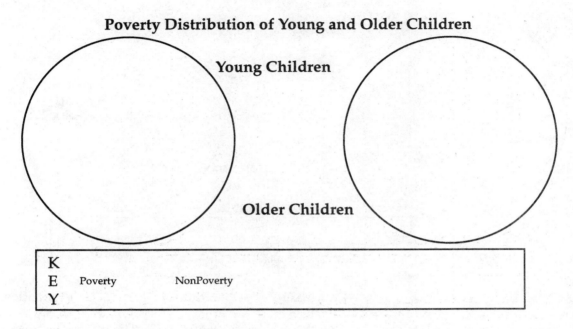

Poverty Distribution of Young and Older Children

Young Children

Older Children

K
E   Poverty   NonPoverty
Y

■ Create a bar chart with side by side bars for young children and older children. For each race/ethnic group, indicate the percentage of young and older children living in poverty.

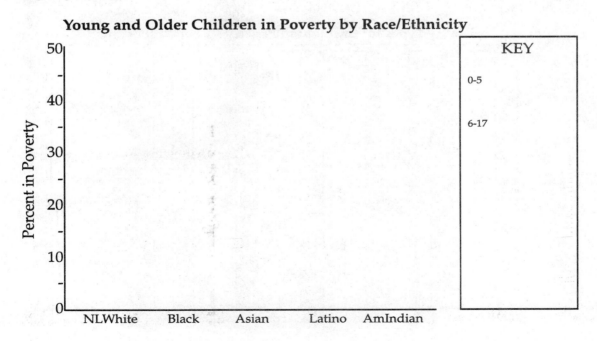

Young and Older Children in Poverty by Race/Ethnicity

KEY

0-5

6-17

Exercise 17 Are children in poverty more prevalent in certain types of families? Focusing on 1990, determine what percentage of all children in poverty live in each family type. For example, what percentage of children in poverty live in married-couple families? *(CHLDPOV9.DAT)*

■ Create two pie charts, one for children in poverty and one for children not in poverty. In each pie, make divisions for family types.

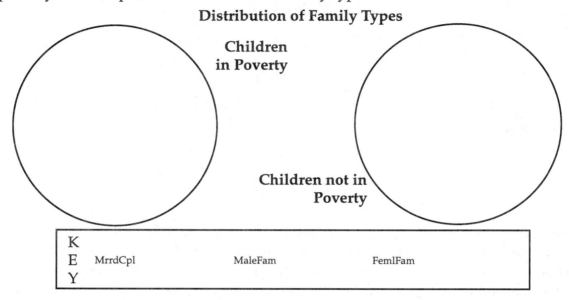

Distribution of Family Types

Children in Poverty

Children not in Poverty

KEY   MrrdCpl   MaleFam   FemlFam

Exercise 18 Do your findings from the previous exercise vary between races/ethnicities? For example, is the family type distribution of white children in poverty different than the family type distribution of Asian children in poverty? Compare the family type distributions of children in and not in poverty for each race/ethnic group. *(CHLDPOV9.DAT)*

■ Create a stacked bar chart with side by side bars for children in and not in poverty. For each race/ethnic group, stack by the family type distribution.

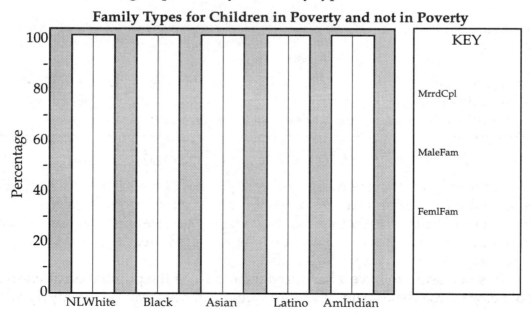

Family Types for Children in Poverty and not in Poverty

KEY

MrrdCpl

MaleFam

FemlFam

NLWhite   Black   Asian   Latino   AmIndian

Exercise 19 Are comfortable income children more prevalent in certain types of families? Focusing on 1990, determine what percentage of all comfortable income children live in each family type.How does this distribution vary from exercise 17? (_CHLDPOV9.DAT_)

■ Create a pie chart for comfortable income children; make divisions for family types.

Distribution of Family Types for Comfortable Income Children

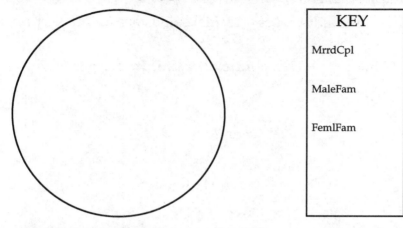

KEY

MrrdCpl

MaleFam

FemlFam

Exercise 20 Do your findings from the previous exercise vary between races/ ethnicities? For example is the family type distribution of comfortable income black children different than the family type distribution of comfortable income Latino children? Using 1990 data, compare the family type distributions of comfortable income children for each race/ethnic group. Compare your findings to those in exercise 19. (_CHLDPOV9.DAT_)

■ On your own, create a stacked bar chart with bars for each race/ethnic group; stack by the family type distribution of comfortable income children.

---

*D*iscussion Questions

1. What are some recent economic and social events that may influence the percentage of children living in poverty?

2. Some critics suggest that child poverty is primarily the result of family structure. Would you agree with this claim? If not, what other social and economic characteristics contribute to the problem of child poverty?

E. Children in School

America's children spend a significant portion of their lives in school. Perhaps more than any other institution, schools are faced with, and expected to address, society's ills. Whether it be child poverty, race relations, gender inequality, or economic competitiveness, policy makers have attempted to use schools to advance and resolve critical social issues.

It is necessary to have a basic understanding of the population attending school before these issues, and other curriculum changes, can be effectively addressed. Who are America's students (ages 0-17)? Are there differences between students who attend public and private schools?

Exercise 21 Focusing on 1990, determine what percentage of school-age children (6-17) are enrolled in public schools. What percentage are enrolled in private schools? (_CHLDSCH9.DAT_)

■ Create a pie chart for children ages 6-17; make divisions for public, private, and not in school.

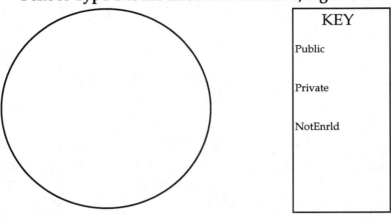

School Type Distribution for Children, Ages 6-17

Exercise 22 Using 1990 data, what percentage of school-age children (6-17) in each race/ethnic group are enrolled in public schools? Private schools? Why might some race/ethnic groups have more students in private schools than others? (_CHLDSCH9.DAT_)

■ Create a stacked bar chart with bars for each race/ethnic group; stack by public, private, and not in school.

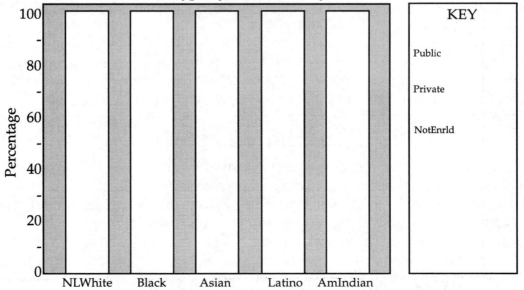

School Type by Race/Ethnicity

Exercise 23 Are poor children less likely to be in private schools? Using 1990 data, examine what percentage of children at or below the poverty line attend private schools. How does this compare to children in families with comfortable incomes? *(CHLDSCH9.DAT)*
■ Create two pie charts, one for children in poverty and one for comfortable income children. In each pie, make divisions for public, private, and not in school.

School Type Distribution of Poor and Comfortable Income Children

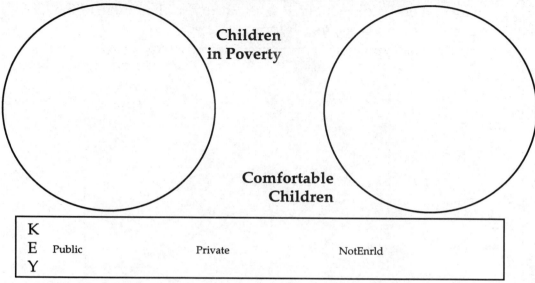

Children in Poverty

Comfortable Children

K
E   Public   Private   NotEnrld
Y

Exercise 24 Using 1990 data, determine what percentage of 18, 21, and 24 year olds in the U.S. are at least high school graduates. Does this vary by race/ethnicity? Describe your findings. *(CHLGRAD9.DAT)*

■ Create a bar chart with side by side bars for ages 18, 21, and 24. For each race/ethnic group, indicate the percentage of high school graduates in each age category.

High School Graduates at Ages 18, 21, and 24

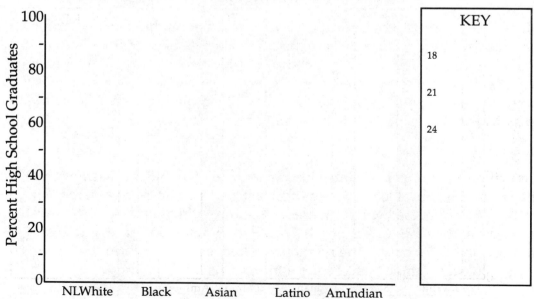

KEY

18

21

24

Discussion Questions

1. As the school-age population becomes more racially/ethnically diverse, many people see an increasing need for *multicultural* education. Unfortunately, there is little consensus about what this actually means. Do you agree that primary and secondary education need to be more *multicultural*? Why or why not?

2. Some policy makers want to make English the *official* language of the U.S. What pros and cons do you see in this type of policy?

THINK tank

1. The U.S. Department of Education is concerned about high school dropouts, and a number of programs have attempted to allocate needed resources to those students least likely to finish high school. If you were in charge of dispersing these limited resources, which students would receive the most funding? Which students appear to have the greatest risk of not completing high school? Identify as specifically as possible the demographic characteristics of high school dropouts.

2. According to the 1990 census, what percentage of all older children (6-17) are proficient in English? Does English proficiency appear to be associated with poverty? Is it connected with whether or not the child attends school — or the type of school they attend? Is English proficiency related to race/ethnicity? Immigrant status? Analyze the relationship between immigration, schooling, poverty, and race/ethnicity. Describe your findings.

Before looking at trends among the elderly, it is necessary to establish exactly who we consider "elderly". Three categories are used to distinguish between the different age groups among the elderly. People between the ages of 65 and 74 are referred to as the "young-old". Individuals whose ages fall between 75 and 84 are in the "old-old" category. Anyone over the age of 85 is part of the "oldest-old" group.

The categorization of the elderly population acknowledges the changes that come with age. While both are considered members of the elderly population, a 65 year-old and an 85 year-old are likely to have very different lives. The 85 year-old will probably have already experienced many of the social and biological changes that accompany old age. The 65 year-old, on the other hand, may only be beginning to experience life changes, such as retirement and physical health challenges.

However, it is important to note that diversity does not rest solely between the age groups. Much variation can also be found within the age groups. Gender, race/ethnicity, marital status, and socioeconomic factors, among others, contribute to the differences between people in the same age group. Historical events also play a significant role by characterizing the different generations. All of these factors should be considered when looking at the elderly and the issues that concern them.

In addition to looking at who the elderly are, it is important to consider why they warrant our attention. Simply put, more Americans are older than ever before. Partially due to modern developments which have increased life expectancies, the elderly have been transformed from a relatively small component into a significant part of the U.S. population. This aging of the population has served as one of the most significant demographic trends in American history. Additionally, as the Baby Boomers

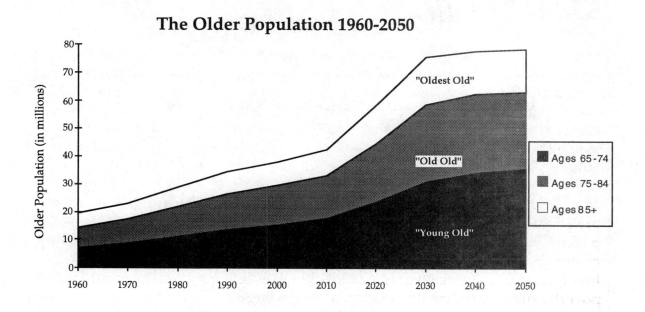

The Older Population 1960-2050

KEY concepts

<u>**Older Population**</u> Persons ages 65 and above who can be further classified according to:

 young-old population ages 65-74
 old-old population ages 75-84
 oldest-old population ages 85 and older

These categories are consistent with recognized stages of changing lifestyles and health.

<u>**Self-Care Limitation**</u> Persons with this limitation report a health condition of six months or more which makes it difficult to take care of personal needs such as dressing, bathing, or getting around inside the house.

<u>**Mobility Limitation**</u> Persons with this limitation report a health condition of six months or more which makes it difficult to go outside the home without assistance.

<u>**Work Disability**</u> Persons with this disability report a health condition of six months or more which limits the level or amount of work they could do at a job.

OTHER concepts

<u>**Race/Ethnicity**</u> (Topic two)
<u>**Marital Status**</u> (Topic five)
<u>**Household Type**</u> (Topic seven)
<u>**Poverty Status**</u> (Topic eight)

<u>**Income Relative to Poverty Cutoff**</u> (Topic eight)
<u>**Labor Force Status**</u> (Topic four)
<u>**Education**</u> (Topic two)

from 1946-1964 grow older, this trend will become even more defined. During the next few decades, boomers long associated with youth will be retiring and joining the ranks of the elderly.

Due to the significant size of the elderly population, few corners of American life and its institutions have been left untouched by the needs, demands, and contributions of the elderly. A significant proportion of all consumers, voters, workers, and home-owners are elderly. Consequently, they have a sizable influence upon the issues that concern such groups. Therefore, a study of the elderly is crucial for a true understanding of American society and the forces that shape it.

In this chapter, we will take a look into the lives of the elderly. We will examine general aspects such as population shifts and trends as well as specific personal factors including family, economic situations, and health. These considerations will pave the way toward an understanding of the elderly and their role in society.

A. Getting Older

Due to increases in life expectancy, both the "old-old" and the "oldest-old" populations have grown and the United States now hosts one of the oldest elderly populations worldwide. Ages that were once considered virtually unattainable are now commonplace in American society.

However, not everyone has been an equal beneficiary of improvements in life expectancy. For example, while the gender gap has narrowed somewhat, women generally live longer lives than men and have made greater gains in life expectancy. While this feminization of aging may seem attractive to women, it can leave them vulnerable to the problems associated with old age such as poverty, widowhood, and institutionalization. Race/ethnicity gaps also exist in life expectancies. Blacks have a life expectancy approximately six to eight years shorter than their white counterparts. Much of this discrepancy stems from the higher mortality rates among younger blacks.

Exercise 1 How has the percentage the elderly comprise of the total population changed over time? Look at men and women separately. What trends do you notice? Are they the same for men and women? (*POP5090.DAT*)

■ Create a line graph with two lines, one for men and one for women. For each year, indicate the percentage of elderly people.

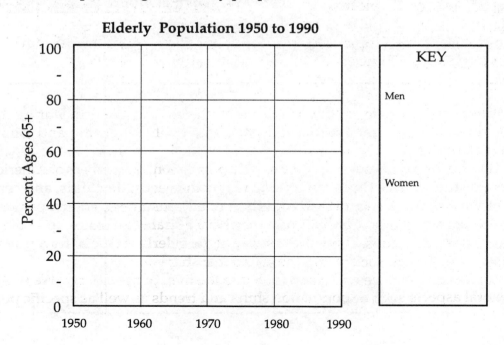

Exercise 2 How has life expectancy changed over time? Examine the distribution of the elderly age groups over time. Are your findings surprising to you? Why or why not? (*ELD5090.DAT*)

■ Create a stacked bar chart with bars for each year; stack by the three elderly age groups.

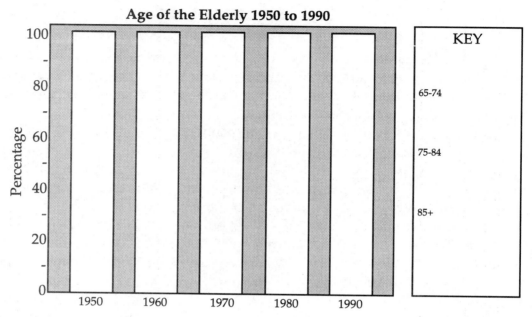

Exercise 3 While keeping the gender gap in mind, look at the number of elderly men and women in each age group. Are women living longer than men? Is this consistent for each age group or does it change? (*ELD5090.DAT*)

■ Create a bar chart with side by side bars for men and women. For each age group, indicate the number of men and women in that group.

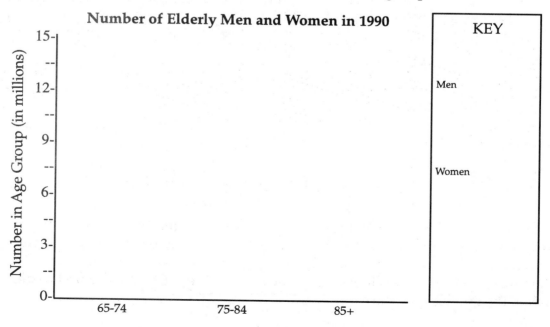

Exercise 4 Now examine how the age distribution of the elderly population differs between race/ethnic groups. Which groups have the greatest percentage of oldest elderly members? (*ELDPOV9.DAT*)

■ Create a stacked bar chart. For each race/ethnic group, stack by elderly age groups.

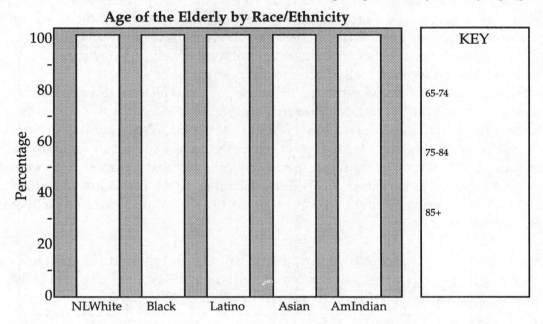

Age of the Elderly by Race/Ethnicity

KEY

65-74

75-84

85+

Discussion Questions

1. We are currently experiencing the "graying" of America. In other words, there is a growing number of elderly citizens. What effects do you think this has on society? Consider aspects of American life that you feel have been affected by the growing number of elderly and cite examples.

2. Keeping gender and race/ethnicity in mind, who has the best chances of living to old age? Why do you think this is so? What factors do you think contribute to the gender and race/ethnicity gaps discussed earlier? Are these gaps inevitable? How could they be narrowed?

B. Marital and Household Changes

Husbands and wives who grow old together run a greater risk of having their marriage end due to the death of a spouse. Since many women marry men who are somewhat older, it is usually the wife who is widowed. Furthermore, due to the gender gap in life expectancy rates, the husbands are usually the first to die even when the spouses are the same age.

When an elderly woman's husband dies, she is unlikely to remarry. While widowed men often continue to live in the community, most widowed women turn to grown children and eventually nursing homes for support. Whereas men draw considerable pensions, women often find themselves slipping into poverty.

Race/ethnicity also plays a significant role in the study of the elderly's marital status. Older black women are less likely to be married than older white women. Similarly, older black women are more likely to have been divorced or to never have been married. Also, black women are more likely to be widowed because black men have especially high mortality rates.

The married couples who manage to survive to old age are not necessarily shielded from the changing trends of marital status. As American divorce rates increased in the 1970s, the elderly were not left untouched. While their numbers are still relatively small, those who do divorce at an old age often cannot enjoy the same resources and lifestyles that they enjoyed while married. Women are more likely to find themselves in financial difficulties than men. As with widowed women, divorced elderly women are likely to turn to their grown children for help whereas men are less likely to do so.

Exercise 5 Begin by looking at the marital status of the elderly in different age groups. Look separately at men and women. What trends emerge as the elderly grow older? Is this the same for men and women? (*ELDPOV9.DAT*)

■ Create a stacked bar chart with side by side bars for men and women. For each age group, stack by marital status.

Marital Status of Elderly Men and Women

Exercise 6 Now compare the marital status of the elderly in each race/ethnic group. What differences do you notice between the race/ethnic groups? (*ELDPOV9.DAT*)

■ Create a stacked bar chart with side by side bars for men and women. For each race/ethnic group, stack by marital status.

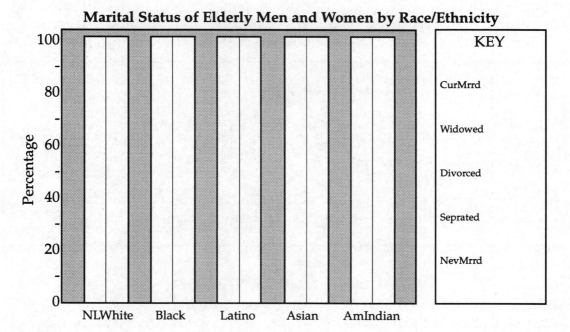

Marital Status of Elderly Men and Women by Race/Ethnicity

Exercise 7 How do household arrangements change as the elderly grow older? Does household type vary between age groups? (*ELDHH9.DAT*)

■ Create a stacked bar chart with bars for each age group (65-74, 75-84, 85+); stack by household type.

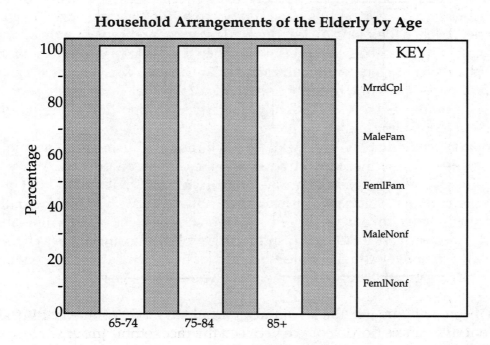

Household Arrangements of the Elderly by Age

Exercise 8 How do household arrangements of the elderly differ between race/ethnic groups? *(ELDHH9.DAT)*

■ On your own, create five pie charts, one for each race/ethnic group. In each pie, indicate percentages of the different household types.

---

*D*iscussion Questions

1. What effect do you think marriage has upon financial stability? How does this differ for men and women? Why do you think this is the case? What could be offered as a possible solution?

2. Recall the household types of different race/ethnic groups. What factors, in your opinion, cause these differences? Social factors? Economic factors?

C. Economic Situation

Mandatory retirement was eradicated in the late 1970s and first half of the 80s. However, this did not spur many dramatic changes in workers' retirement plans. Generally speaking, workers chose to retire sooner than later. In other words, retirement became merely a matter of timing. With the support of social security, private pensions, and savings, many look forward to filling their days with activities other than those done at the office.

However, retirement is not always a voluntary release from the work force. Health problems often restrict individuals' abilities to fulfill their responsibilities and they are forced to leave their jobs. Other times, some accept retirement only because they fear losing their jobs. These workers, when they choose to find alternate work, are often further discouraged by difficulties in finding a new job. Even when they are able to find part-time work, these jobs often do not pay well and do not offer the benefits that the elderly need. Consequently, a significant proportion of the elderly are left vulnerable to financial distress.

The economic situations of the elderly can be assessed using the "Income Relative to Poverty Cutoff" measure introduced in the chapter on poverty. While most of the elderly are in the middle income category, a significant number fall into the poor and near-poor categories. The near-poor, however, are at high risk of falling below the poverty line with any minor setback. Women are particularly vulnerable to poverty and near-poverty. Education and race/ethnicity also affect one's chances of staying above the poverty line. While a majority of those who are poor are white, racial minorities are greatly over-represented among the poor elderly.

Exercise 9 Begin by looking at the labor force participation of the elderly in 1990. Look at men and women separately. How does labor force participation change as age increases? Is this the same for men and women? *(ELDEMP9.DAT)*

■ Create two bar charts, one for men and one for women. In each chart, indicate the percentage of each age group in the labor force. (Hint: Combine Unempd, EmpFull, and EmpPart; See "Key Concepts" in the Labor Force chapter.)

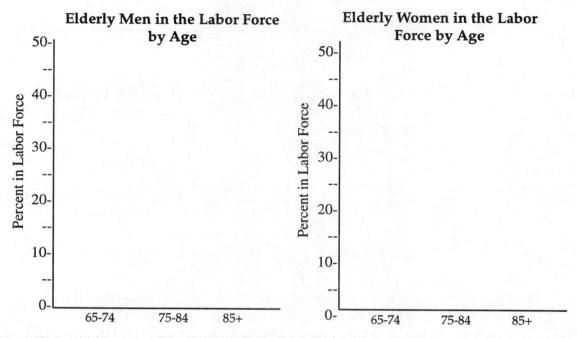

Exercise 10 Compare the income distributions of elderly men and women (use the "Income Relative to Poverty Cutoff" measure). How do men and women differ in terms of poverty? In terms of being comfortable? (*ELDPOV9.DAT*)

■ Create two pie charts, one for elderly men and one for elderly women. In each pie, show the income distribution.

Income Distribution of Men and Women Ages 65+

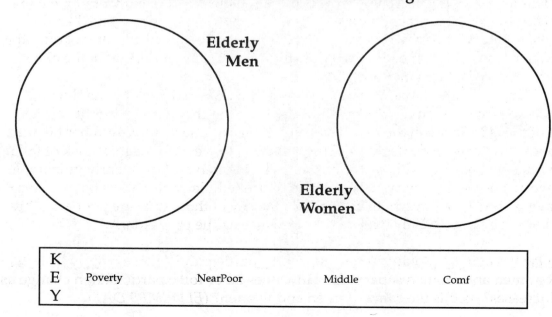

Exercise 11 Does poverty among the elderly vary by race/ethnicity? Which groups have the highest poverty levels? The lowest? Why do you think this is so? (*ELDPOV9.DAT*)

■ Create a stacked bar chart. For each race/ethnic group, stack by income status.

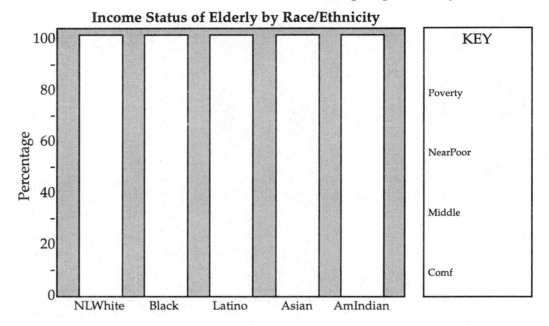

Income Status of Elderly by Race/Ethnicity

Exercise 12 Consider the relationship between marital status and poverty. Does the relationship differ between men and women? If so, how? (*ELDPOV9.DAT*)

■ Create a stacked bar chart with side by side bars for men and women. For each marital category, stack by income status.

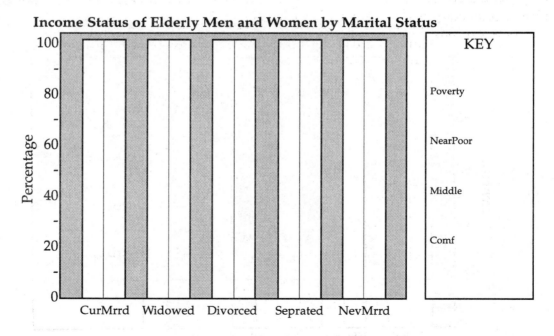

Income Status of Elderly Men and Women by Marital Status

Exercise 13 Examine the educational attainment of the elderly over time. How has it changed? Why do you think this is so? (*EDUC5090.DAT*)

■ Create a stacked bar chart with side by side bars for men and women. For each year, stack by educational attainment.

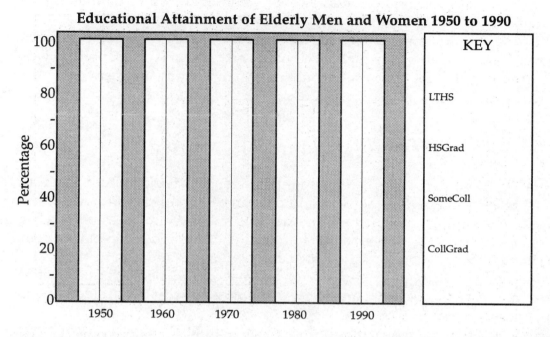

Educational Attainment of Elderly Men and Women 1950 to 1990

Exercise 14 Now look at the elderly's poverty rates across different educational attainment categories. Are those with higher education levels less likely to be in poverty? Why do you think this is the case? (*ELDEMP9.DAT*)

■ Create a stacked bar chart with side by side bars for men and women. For each educational category, stack by income status.

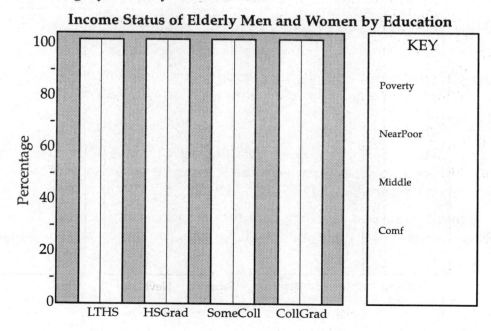

Income Status of Elderly Men and Women by Education

𝒟. Health and Disability

The greater part of the elderly population report having at least one recurring health ailment. Because some conditions can be life-threatening, such as heart disease, the elderly are often in need of extensive medical care. Less serious health problems, such as arthritis, pose a threat to quality of life and independent living. These ailments often require assistance in daily living. In addition to physical disorders, many elderly people suffer from mental conditions, such as Alzheimer's disease. Alzheimer's disease is one of the major reasons why many elderly people are institutionalized.

Those who need assistance, but are not institutionalized, must turn to other sources for help. Such sources often include spouses, family members, and friends. Other times, special services and agencies are called upon to provide the help that is needed. However, these services are expensive and not available in all communities. Therefore, those closer to the poverty level are sometimes not able to receive the professional assistance they need.

Exercise 15 Does income level have an effect upon the presence of self-care limitations among the elderly? How? Why do you think this is so? (*ELDDSAB9.DAT*)

■ On your own, create four pie charts, one for each income status category. In each pie, indicate the percentage of those who have self-care limitations and those who do not.

Exercise 16 Do the elderly's self-care limitations vary by race/ethnicity? What differences do you notice between race/ethnic groups? What is a possible explanation for these differences? (*ELDDSAB9.DAT*)

■ On your own, create five pie charts, one for each race/ethnic group. For each pie, indicate the percentage of those who have self-care limitations and those who do not.

Exercise 17 Consider the self-care limitations among the elderly. Who has such limitations? Who does not? How do men and women differ in respect to self-care limitations? (*ELDDSAB9.DAT*)

■ Create a stacked bar chart with side by side bars for men and women. For each age group, stack by those who have self-care limitations and those who do not.

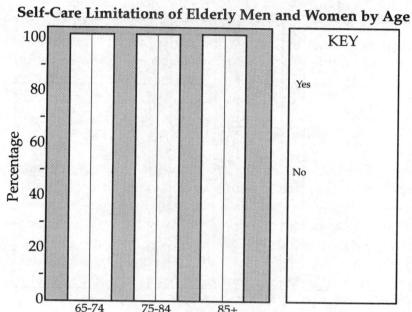

Self-Care Limitations of Elderly Men and Women by Age

Exercise 18 Now consider mobility limitations among the elderly. How do men and women differ in respect to mobility limitations? What differences do you notice between age groups? (*ELDDSAB9.DAT*)

■ Create a stacked bar chart with side by side bars for men and women. For each age group, stack by mobility limitation status.

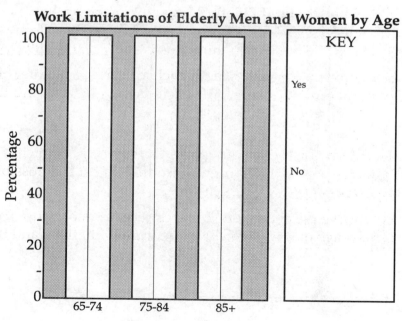

Work Limitations of Elderly Men and Women by Age

Exercise 19 How do mobility limitations among the elderly vary by income status? Do you note any significant relationships? *(ELDDSAB9.DAT)*

■ Create a stacked bar chart, with bars for each of the "Income Relative to Poverty Cutoff" categories; stack by mobility limitation status.

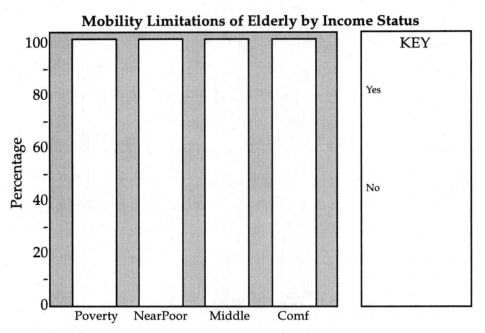

Mobility Limitations of Elderly by Income Status

Exercise 20 Now take a look at work limitations, often a result of the two previous limitations you have examined. How do men and women differ in respect to work limitations? *(ELDDSAB9.DAT)*

■ Create a stacked bar chart with side by side bars for men and women. For each age group, stack by work limitation status.

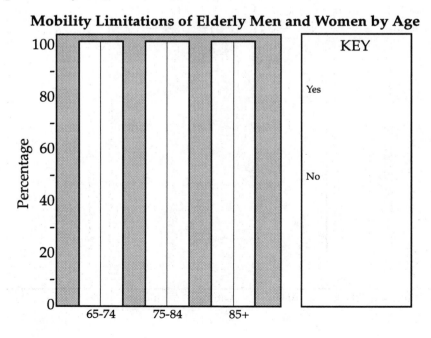

Mobility Limitations of Elderly Men and Women by Age

Exercise 21 How do work limitations among the elderly vary by income status? (*ELDDSAB9.DAT*)

■ Create a stacked bar chart, with bars for each of the "Income Relative to Poverty Cutoff" categories; stack by work limitation status.

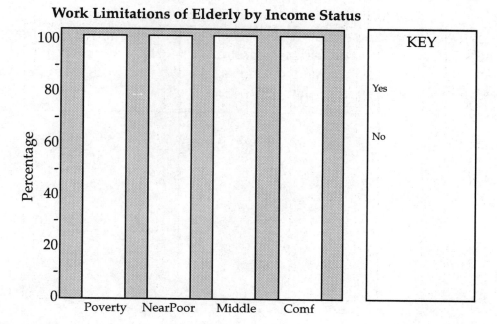

Work Limitations of Elderly by Income Status

Exercise 22 How do work limitations among the elderly vary by race/ethnicity? (*ELDDSAB9.DAT*)

■ Create a stacked bar chart with bars for each race/ethnic group; stack by work limitation status.

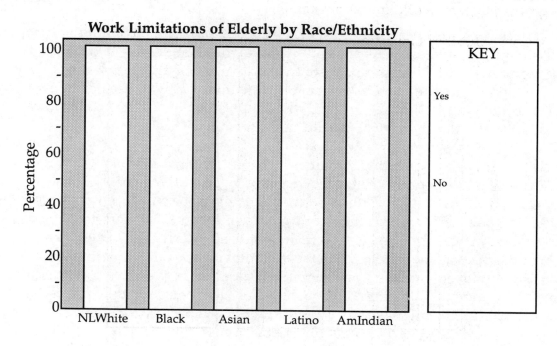

Work Limitations of Elderly by Race/Ethnicity

Discussion Questions

1. If we had looked at the limitations of the elderly since 1950, do you think you would have noticed a change over time? What kind of change? Provide a possible explanation. Do you think it will continue to change in the future?

2. Looking at limitation distributions among the different income and race/ethnicity categories, do you feel any parallels exist? If so, how? What, in your opinion, do you think this says about our national health policies in the United States? How could things be changed?

THINK tank

1. Describe where you would most likely find elderly Latinos. Would you be just as likely to find older Asians and Blacks in the same setting? Does the location for any of these groups seem to vary by region? Given that the elderly population is growing, and that the elderly are more likely to use health care services than most other age groups, what implications does the geographical location of the elderly have for health care providers?

2. The elderly are living longer than ever before, but are they also working longer? What percentage of elderly were employed in 1950? How does this compare with 1990? Were elderly women as likely to be employed between 1950-1990 as elderly men? Considering employment in 1990, were employed elderly women as likely to work as many hours as elderly men? Does race/ethnicity appear to be a factor? Do these employment characteristics seem consistent with elderly work and mobility limitations? Given that the elderly are living longer, do you think the official retirement age should be pushed back as well? How might such a change affect Social Security, Medicare, or other programs designed to assist older Americans?

SECTION III
References and Resources

Investigating Change in American Society

REFERENCES FOR TOPICS

topic one: Population Structure: Cohorts, Ages and Change

Bouvier, Leon F., and Carol J. De Vita. "The Baby Boom— Entering Midlife." *Population Reference Bureau Bulletin* 46, no. 3, 1991.

Crispell, Diane. "Generations 2025." *American Demographics* 17, no.1, January 1995: 4.

De Vita, Carol J. "United States at Mid-Decade." *Population Reference Bureau Bulletin* 50, no. 4, 1996.

Dunn, William. *The Baby Bust: A Generation Comes of Age*. Ithaca, NY: American Demographics, 1993.

Easterlin, Richard A. *Birth and Fortune: The Impact of Numbers on Personal Welfare*. Chicago, IL: The University of Chicago Press, 1987.

Frey, William H. "Metropolitan America: Beyond the Transition." *Population Reference Bureau Bulletin* 45, no. 2, 1990.

Frey, William H. "The New Geography of Population Shifts." *State of the Union, America in the 1990s, Volume Two: Social Trends* (pp.271-336). Edited by Reynolds Farley. New York, NY: Russell Sage Foundation, 1995.

Gober, Patricia. "Americans on the Move." *Population Reference Bureau Bulletin* 48, no. 3, 1993.

Haub, Carl. "Understanding Population Projections." *Population Bulletin* 42, no.4, December 1987.

Howe, Neil, and Bill Strauss. *13th Gen*. New York, NY: Vintage Books, 1993.

Johnson, Kenneth M., and Calvin L. Beale. "The Rural Rebound Revisited." *American Demographics* 17, no. 7, July 1995: 46-54.

Jones, Landon Y. "Great Expectations and the Baby Boom Generation." New York: Coward, McCann, and Geoghegan, 1980.

Longino, Jr. Charles F. *Retirement Migration in America*. Houston, TX: Vacation Publications, 1995.

Mitchell, Susan. "The Next Baby Boom." *American Demographics* 17, no. 10, October 1995: 22-31.

Murdock, Steve H. *America Challenged: Population Change and the Future of the United States*. Boulder, CO: Westview Press, 1995.

Ritchie, Karen. *Marketing to Generation X*. New York: Lexington Books, 1995.

Russell, Cheryl. "The Baby Boom Turns 50." *American Demographics* 17, no. 12, December 1995: 22-33.

Russell, Cheryl. *The Master Trend: How the Baby Boom Generation is Remaking America*. New York, NY: Plenum Publications, 1993.

Weeks, John R. "Age and Sex Structure." *Population: An Introduction to Concepts and Issues, 6th Edition*, Chapter 8. Belmont, CA: Wadsworth, 1996.

——————— *topic two:* **Race and Ethnic Inequality** ———————

Allen, Walter R., and Joseph O. Jewell. "African American Education Since An American Dilemma." *Daedalus: Journal of the American Academy of Arts and Sciences* 124, no. 1, December 1995: 77-100.

Baringer, Hebert, and Robert W. Gardner. *Asians and Pacific Islanders in the United States*. New York, NY: Russell Sage Foundation, 1993.

Bean, Frank D., and Marta Tienda. *The Hispanic Population of the United States*. New York, NY: Russell Sage Foundation, 1987.

Dentler, Robert A. "The Political Sociology of the African American Situation: Gunnar Myrdal's Era and Today." *Daedalus: Journal of the American Academy of Arts and Sciences* 124, no. 1, December 1995: 15-36.

De Vita, Carol J. "United States at Mid-Decade." *Population Reference Bureau Bulletin* 50, no.4, 1996.

Farley, Reynolds, and Walter R. Allen. *The Color Line and the Quality of Life in America*. New York, NY: Russell Sage Foundation, 1987.

Farley, Reynolds. *The New American Reality*. New York, NY: Russell Sage Foundation, 1996.

Frey, William H. "The New Geography of Population Shifts." *State of the Union, America in the 1990s, Volume Two: Social Trends*, pp.271-336. Edited by Reynolds Farley. New York, NY: Russell Sage Foundation, 1995.

Harrison, Roderick J., and Claudette Bennett. "Racial and Ethnic Diversity." *State of the Union, America in the 1990s, Volume Two: Social Trends* (pp. 141-201). Edited by Reynolds Farley. New York, NY: Russell Sage Foundation, 1995.

Kitano, Harry H. L. *Asian Americans: Emerging Minorities*. Englewood Cliffs, NJ: Prentice Hall, 1988.

Lieberson, Stanley, and Mary C. Waters. *From Many Strands: Ethnic and Racial Groups in Contemporary America*. New York, NY: Russell Sage Foundation, 1987.

Mare, Robert D. "Changes in Educational Attainment and School Enrollment." *State of the Union, America in the 1990s, Volume One: Economic Trends* (pp. 155-213). Edited by Reynolds Farley. New York, NY: Russell Sage Foundation, 1995.

O'Hare, William P., Kelvin M. Polland, Taynia L. Mann, and Mary M. Kent. "African Americans in the 1990s." *Population Reference Bureau Bulletin* 46, no. 1, 1991.

O'Hare, William P., and Judy C. Felt. "Asian Americans: America's Fastest Growing Minority Group." *Population Reference Bureau Policy Report* 19, 1991.

Ong, Paul. *The State of Asian Pacific America's Economic Diversity, Issues and Policies,* Los Angeles, CA: LEAP Asian Pacific American Public Policy Institute and Asian American Studies Center, 1995.

Rose, Stephen. *Social Stratification in the United States: The American Profile Poster Revised and Expanded.* New York, NY: The New Press, 1992.

Valdivieso, Rafael, and Carey Davis. "US Hispanics: Challenging Issues for the 1990s." *Population Reference Bureau Policy Reports* 17, 1988.

Weeks, John R. "Population Characteristics and Life Chances." *Population: An Introduction to Concepts and Issues, 6th Edition,* Chapter 8. Belmont, CA: Wadsworth, 1996.

———— *topic three:* Immigrant Assimilation ————

Chiswick, Barry R., and Teresa Sullivan. "The New Immigrants." *State of the Union, America in the 1990s, Volume Two: Social Trends* (pp. 211-70). Edited by Reynolds Farley. New York, NY: Russell Sage Foundation, 1995.

Davidson, Mariam. "Second-Class Refugees: Persecuted Women Are Denied Asylum." *The Progressive* 58, May 1994: 22-5.

De Vita, Carol J. "United States at Mid-Decade." *Population Reference Bureau Bulletin* 50, no.4, 1996: 2-48.

Edmonston, Barry and Jeffrey S. Passel. *Immigration and Ethnicity: The Interaction of America's Newest Arrivals.* Washington DC: The Urban Institute Press, 1994.

Fix, Michael, and Jeffrey S. Passel. *Immigration and Immigrants: Setting the Record Straight.* Washington DC: The Urban Institute Press, 1994.

Frey, William H. "The New Geography of Population Shifts." *State of the Union, America in the 1990s, Volume Two: Social Trends* (pp.271-336). Edited by Reynolds Farley. New York, NY: Russell Sage Foundation, 1995.

Jenson, Leif, and Yoshimi Chitose. "Today's Second Generation: Evidence from the 1990 US Census." *International Migration Review* 28, no. 4, December 1994: 714-35.

Martin, Philip, and Elizabeth Midgley. "Immigration to the US: Journey to an Uncertain Destination." *Population Reference Bureau Bulletin* 49, no. 2, 1994.

Ong Hing, Bill, and Ronald Lee. "Reframing the Immigration Debate." Los Angeles, CA: Asian Pacific American Public Policy Institute and UCLA Asian American Studies Center, 1996.

Pedraza, Silvia and Reuben Rumbaut. *Origins and Destinies: Immigration, Race, and Ethnicity.* Belmont, CA: Wadsworth Publishing Company, 1996.

Portes, Alejandro. *The New Second Generation.* New York: Russell Sage, 1996.

Waldinger, Roger., and Mehdi Bozorgemehr. *Ethnic Los Angeles.* New York, NY: Russell Sage, 1996.

Bianchi, Suzanne M. and Daphne G. Spain. *American Women in Transition.* New York, NY: Russell Sage Foundation, 1986.

Braus, Patricia. "How Women will Change Medicine." *American Demographics* 16, no. 11, November 1994: 40-47.

De Vita, Carol J. "United States at Mid-Decade." *Population Reference Bureau Bulletin* 50, no.4, 1996.

Dortch, Sannon. "For This I Waited?" *American Demographics* 16, no. 10, October 1994: 14-16.

Farley, Reynolds, and Walter R. Allen. *The Color Line and the Quality of Life in America.* New York, NY: Russell Sage Foundation, 1987.

Hodson, Randy, and Teresa A. Sullivan. *The Social Organization of Work.* Belmont, CA: Wadsworth Publishing Company, 1990.

Kasarda, John D. "Industrial Restructuring and the Changing Location of Jobs." *State of the Union, America in the 1990's, Volume One: Economic Trends* (pp. 215-67). Edited by Reynolds Farley. New York, NY: Russell Sage Foundation, 1995.

Kessler-Harris, Alice. *Women Have Always Worked: A Historical Overview.* New York, NY: The Feminist Press, 1981.

Levy, Frank. *Dollars and Dreams: The Changing American Income Distribution.* New York, NY: Russell Sage Foundation, 1987.

Levy, Frank. "Incomes and Income Inequality." *State of the Union, America in the 1990s, Volume One: Economic Trends* (pp. 1-57). Edited by Reynolds Farley. New York, NY: Russell Sage Foundation, 1995.

O'Connell, Martin, and David E. Bloom. "Juggling Jobs and Babies: America's Child Care Challenge." *Population Reference Bureau Policy Report* 12, 1987.

Reich, Robert B. *The Work of Nations: Preparing Ourselves for 21st Century Capitalism.* New York, NY: Vintage Books, 1991.

Sacks, Karen B. and Dorothy Remy. *My Troubles Are Going To Have Trouble with Me: Everyday Trials and Triumphs of Women Workers* (pp. 127-141). Rutgers, NJ: Rutgers University Press, 1984.

Spain, Daphne G., and Suzanne M. Bianchi. *Balancing Act: Marriage, Motherhood and Employment Among American Women.* New York, NY: Russell Sage, 1996.

Sullivan, Teresa A. "The Cashier Complex and the Changing American Labor Force" edited by Dennis L. Peck and J. Selwyn Hollingsworth. *Demographic Change and Structural Change.* Westport, Conneticut: Greenwood Press, 1996.

Wetzel, James R. "Labor Force, Unemployment, and Earnings." *State of the Union, America in the 1990s, Volume 1: Economic Trends* (pp. 59-105). Edited by Reynolds Farley. New York, NY: Russell Sage Foundation, 1995.

Williams, Christine L. *Gender Differences at Work: Women and Men in Nontraditional Occupations.* Berkeley, CA: University of California Press, 1989.

— *topic five:* Marriage, Divorce, Cohabitation, and Childbearing——

Bianchi, Suzanne M. and Daphne G. Spain. *American Women in Transition.* New York, NY: Russell Sage Foundation, 1986.

Cherlin, Andrew J. *Marriage, Divorce, Remarriage.* Cambridge, MA: Harvard University Press, 1992.

Crispell, Diane. "Dual-Earner Diversity." *American Demographics* 17, July 1995: 32-37.

Da Vanzo, Julie, and Rahman Omar M. *American Families: Trends and Correlates.* Santa Monica, CA: Rand Corporation, 1994.

De Vita, Carol J. "United States at Mid-Decade." *Population Reference Bureau Bulletin* 50, no.4, 1996.

Furstenburg, Jr. Frank F., and Andrew J. Cherlin. "Divorced Families: What Happens to Children When Parents Part." Cambridge, MA: Harvard University Press, 1991.

Furstenberg, Jr. Frank F. "The Future of Marriage." *American Demographics* 18, June 1996: 34-40.

Goldscheider, Calvin and Frances. "Leaving and Returning Home in 20th Century America." *Population Reference Bureau Bulletin* 48, no. 4, 1994.

Spain, Daphne G., and Suzanne M. Bianchi. *Balancing Act: Marriage, Motherhood and Employment Among American Women.* New York, NY: Russell Sage, 1996.

Waite, Linda J. "Does Marriage Matter?" *Demography* 32, no. 4, November 1995: 483-507.

Weeks, John R. "Population Growth, Women and the Family." *Population: An Introduction to Concepts and Issues, 6th Edition* (Chapter 10). Belmont, CA: Wadsworth, 1996.

—————————— *topic six:* Gender Inequality ——————————

American Demographics. *Women Change Places.* Ithaca, NY: American Demographics, 1993.

Bianchi, Suzanne M. "Changing Economic Roles of Women and Men." *State of the Union: America in the 1990s, Volume One: Economic Trends* (pp. 107-54). Edited by Reynolds Farley. New York, NY: Russell Sage Foundation, 1995.

Bianchi, Suzanne M. and Daphne G. Spain. *American Women in Transition.* New York, NY: Russell Sage Foundation, 1986.

Cockburn, Cynthia. *In the Way of Women: Men's Resistance to Sex Equality in Organizations.* Ithaca, NY: ILR Press, 1991.

De Vita, Carol J. "United States at Mid-Decade." *Population Reference Bureau Bulletin* 50, no.4, 1996.

Levy, Frank. "Incomes and Income Inequality." *State of the Union, America in the 1990s, Volume One: Economic Trends* (pp. 1-57). Edited by Reynolds Farley. New York, NY: Russell Sage Foundation, 1995.

Rollins, Judith. *Between Women: Domestics and Their Employers*. Philadelphia, PA: Temple University Press, 1985.

Russell, Cheryl, "Glass Ceilings Can Break." *American Demographics* 17, no. 11, November 1995: 8.

Sacks, Karen B. and Dorothy Remy. *My Troubles Are Going To Have Trouble with Me: Everyday Trials and Triumphs of Women Workers*, Rutgers, NJ: Rutgers University Press, 1984.

Spain, Daphne G., and Suzanne M. Bianchi. *Balancing Act: Marriage, Motherhood and Employment Among American Women*. New York, NY: Russell Sage, 1996.

Sweet, James A., and Larry Bumpass. *American Families and Households*. New York, NY: Russell Sage, 1987.

Wetzel, James R. "Labor Force, Unemployment, and Earnings." *State of the Union, America in the 1990s, Volume 1: Economic Trends* (pp. 59-105). Edited by Reynolds Farley. New York, NY: Russell Sage Foundation, 1995.

Williams, Christine L. *Gender Differences at Work: Women and Men in Nontraditional Occupations*. Berkeley, CA: University of California Press, 1989.

——————————— *topic seven:* **Households and Families**———————————

Ahlburg, Dennis A., and Carol J. De Vita. "New Realities of the American Family." *Population Reference Bureau Bulletin* 47, no. 2, 1992.

Da Vanzo, Julie, and Omar M. Rahman. *American Families: Trends and Correlates*. Santa Monica, CA: Rand Corporation, 1994.

De Vita, Carol J. "United States at Mid-Decade." *Population Reference Bureau Bulletin* 50, no.4, 1996.

Furstenburg, Jr. Frank F., and Andrew J. Cherlin. "Divorced Families: What Happens to Children When Parents Part." Cambridge, MA: Harvard University Press, 1991.

Goldscheider, Calvin and Frances. "Leaving and Returning Home in 20th Century America." *Population Reference Bureau Bulletin* 48, no. 4, 1994.

Goldscheider, Frances, and Linda Waite. *New Families, No Families?* Berkeley, CA: University of California Berkeley Press, 1991.

Levy, Frank, and Richard C. Michel. *Economic Future of American Families: Income and Wealth Trends*. Washington DC: The Urban Institute Press, 1991.

McLanahan, Sara, and Lynn Casper. "Growing Diversity and Inequality in the American Family." *State of the Union, America in the 1990s, Volume Two: Social Trends* (pp. 1-45). Edited by Reynolds Farley. New York, NY: Russell Sage Foundation, 1995.

Myers, Dowell, and Jennifer R. Wolch. "The Polarization of Housing Status." *State of the Union, America in the 1990s, Volume One: Economic Trends* (pp. 269-334). Edited by Reynolds Farley New York, NY: Russell Sage Foundation, 1995.

Moore, Kristin. *Report to Congress on Out-of-Wedlock Childbearing (Executive Summer)*. Washington DC: U.S. Department of Health and Human Services (DHHS Pub. no. (PHS) 95-1252-1), 1995.

O'Connell, Martin. "Where's Papa? Fathers Role in Childcare." *Population Trends and Public Policy*, no. 20, September 1993.

Sweet, James A., and Larry Bumpass. *American Families and Households*. New York, NY: Russell Sage, 1987.

Whitehead, Barbara Defoe. "Dan Quayle Was Right." *The Atlantic* 271, no. 4, 1993: 47-84.

—————————————— *topic eight:* Poverty ——————————————

Danziger, Sheldon H., and Daniel H. Weinberg. "The Historical Record: Trends in Family Income, Inequality, and Poverty." *Confronting Poverty: Prescriptions for Change* (pp. 1-17). Edited by Sheldon H. Danziger, Gary D. Sandefur, and Daniel H. Weinberg. Russell Sage Foundation: New York, NY, 1994.

Danziger, Sheldon H. and Peter Gottschalk. *America Unequal*. Cambridge, MA: Harvard University Press, 1995.

De Vita, Carol J. "United States at Mid-Decade." *Population Reference Bureau Bulletin* 50, no.4, 1996.

Frey, William H. and Elaine L. Fielding. "Changing Urban Populations, Racial Polarization, and Poverty Concentration." *Cityscape* Vol. 1, no. 2, 1995: 1-66.

Jennings, James. *Understanding the Nature of Poverty in Urban America*. Westport, CT: Praeger, 1994.

O'Hare, William P. "Poverty in America: Trends and New Patterns." *Population Reference Bureau Bulletin* 51, no. 2, 1996.

Treas, Judith, and Ramon Torrecilha. "The Older Population." *State of the Union, America in the 1990s, Volume Two: Social Trends* (pp. 47-92). Edited by Reynolds Farley. New York, NY: Russell Sage Foundation, 1995.

—————————————— *topic nine:* Children ——————————————

Annie E. Casey Foundation. *Kids Count Data Book 1995: State Profiles of Child Well-Being*. Baltimore, MD: Annie E. Casey Foundation, 1995.

Baker, Linda. "Day-Care Disgrace." *The Progressive* 58, June 1994: 26-7.

Bianchi, Suzanne M. "America's Children: Mixed Prospects." *Population Reference Bureau Bulletin* 45, no. 1, 1990.

Children's Defense Fund. *The State of America's Children Yearbook* Washington DC: Children's Defense Fund, 1996.

De Vita, Carol J. "United States at Mid-Decade." *Population Reference Bureau Bulletin* 50, no.4, 1996.

Furstenburg, Jr. Frank F., and Andrew J. Cherlin. "Divorced Families: What Happens to Children When Parents Part." Cambridge, MA: Harvard University Press, 1991.

Hernandez, Donald J. *America's Children: Resources for Family, Government, and the Economy*. New York, NY: Russell Sage Foundation, 1992.

Hogan, Dennis P., and Daniel T. Lichter. "Children and Youth: Living Arrangements and Welfare." *State of the Union, America in the 1990s* (pp. 93-140). Edited by Reynolds Farley. New York, NY: Russell Sage Foundation, 1995.

Jenson, Leif, and Yoshimi Chitose. "Today's Second Generation: Evidence from the 1990 US Census." *International Migration Review* 28, no. 4, December 1994: 714-35.

Jones, Rebecca M. "The Price of Welfare Dependency." *Social Work* 40, no. 4, July 1995: 496-505.

McLanahan, Sara and Gary Sandefur. *Growing Up with a Single Parent*. Cambridge, MA: Harvard University Press, 1994.

O'Connell, Martin. "Where's Papa? Fathers' Role in Child Care." *Population Reference Bureau Policy Report* 20, 1993.

O'Hare, William P. "Children in Distressed Neighborhoods." American Demographics 16, no. 11, November 1994: 19.

Tracy, Elizabeth M. "Maternal Substance Abuse: Protecting the Child, Preserving the Family." *Social Work* 39, no. 5, September 1994: 534-40.

—————————————— *topic ten:* **The Older Population** ——————————————

De Vita, Carol J. "United States at Mid-Decade." *Population Reference Bureau Bulletin* 50, no.4, 1996.

Frey, William H. "Metropolitan Redistribution of the U.S. Elderly, 1960-70, 1970-80 and 1980-90" in Andrei Rogeis (ed) *Elderly Migration and Population Redistribution: A Comparative Perspective*. London: Belhaven, 1992: 123-142.

Hobbs, Frank B. with Damon, Bonnie L. *65+ in the United States*, U. S. Bureau of the Census, Current Population Report, Special Studies, Washington DC: U.S. Government: 23-190, 1996.

Ozawa, Martha N. "The Economic Status of Vulnerable Older Women." *Social Work* 40, no. 3, May 1995: 323-41.

Siegel, Jacob S. *A Generation of Change: A Profile of America's Older Population*. New York, NY: Russell Sage Foundation, 1991.

Soldo, Beth J., and Emily M. Agree. "America's Elderly." *Population Reference Bureau Bulletin* 43, no. 3, 1988.

Treas, Judith. "Older Americans in the 1990s and Beyond." *Population Reference Bureau Bulletin* 50, no. 2, May 1995.

Treas, Judith, and Ramon Torrecilha. "The Older Population." *State of the Union, America in the 1990s, Volume Two: Social Trends*, 47-92. Edited by Reynolds Farley. New York, NY: Russell Sage Foundation, 1995.

Weeks, John R. "Population Growth and Aging." *Population: An Introduction to Concepts and Issues, 6th Edition*. Belmont, CA: Wadsworth, 1996.

WHAT IS THE CENSUS?

All of the population, household and housing data used in this book's investigations are obtained from U.S. censuses. Taken every ten years, the census is the most comprehensive source for these statistics because it is charged by the U.S. Constitution to count every resident of the United States for the immediate purpose of apportioning congressional representatives every ten years. The first census was conducted in 1790. The year 2000 census marks the 22nd census in the nation's history.

The 1990 Census counted over 248 million people and 106 million housing units, receiving most responses within two weeks of Census Day — April 1, 1990. Answering the census is required by law and a successful census involves a widespread public awareness of this obligation. An example of the kind of promotional material that gets distributed shortly before Census Day is shown below.

Although the official purpose of the Census is to count everyone for the purpose of congressional representation, its broader significance involves more than just "counting people". Due to the vast variety of social,

EVERYONE COUNTS!

Answer the Census

U.S. Department of Commerce BUREAU OF THE CENSUS

economic and housing information it collects, it is useful for planning by governments at all levels, and provides a ten-year "benchmark" of all aspects of the nation's population.

The wide range of statistics collected by the decennial census is especially useful in social science research. This is because this information is collected for a large number of people. This means that detailed social and economic information can be gathered for tiny population subgroups and small geographic areas. Unlike many small surveys, the census information is rarely limited by having "too few observations" to be statistically representative.

Recent censuses have typically issued two different types of census forms for people to fill out. One form called the "short form" questionnaire includes questions that are asked of residents in all households. In 1990, short-form population questions included age, gender, race, Hispanic origin, marital status and household relationship. Ten housing questions were also asked. One out of every six households received a "long-form" questionnaire which not only included the short-form items, but also a larger battery of questions related to a person's social characteristics (such as educa-

Stand up and be counted.

April 1.
Answer the census.

tion and English language proficiency), economic characteristics (such as occupation and labor force status), and housing. Although the "long-form" items were only asked of some residents, the sample is big enough to yield accurate estimates for the total population, even for small population groups and areas.

Another reason that census information is valuable for social scientists is because it can provide an over-time record of change in the nation's population, household and housing characteristics. The census is a valuable data source for studying social and economic change for the country as a whole, and for smaller groups within the U.S., because it is taken every ten years and gathers a wide range of information.

This ability to study change is especially important for the investigations in this book which focus on the decades between 1950 and 1990. Several items in the

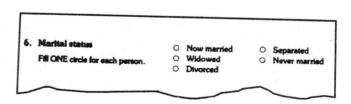

6. Marital status
Fill ONE circle for each person.
   O Now married   O Separated
   O Widowed   O Never married
   O Divorced

1990 census are repeated in each census over this period — making it possible to investigate important societal changes of population attributes, such as education attainment, occupation, earnings, marital status, household relationship, and many others. Some items change across censuses, both in the way a question is asked and the way the responses are coded — consistent with changing times. For example, occupations are categorized differently in different years (as jobs tended to change in the U.S.). Still, it is possible to reallocate these categories into consistent broad categories used in this book's 1950 to 1990 analysis. Other items, such as marital status (depicted above as it appeared on the 1990 census questionnaire), has appeared in almost the same form in many earlier censuses. In fact, the question on marital status has appeared on every U.S. census since 1880.

The census distributes its results in the form of tables and computerized data files that are available in many formats. Published volumes are available for all U.S. censuses and can be found, for the most recent censuses, in most college, university and large public libraries. In the mid-1990s, the Census Bureau began an effort to make census tables and other information available electronically over the Internet. This is discussed

What's in the Census for Me?

No one gets paid for answering the census, but it pays off for everyone.

The information your answers provide helps your community leaders decide where to put day care centers, schools, hospitals, and many more services. And, the census is used to determine how many seats your state has in the U.S. House of Representatives so your voice is heard where it counts the most.

Answer the Census. It Counts for More Than You Think!

in the next section on "Resources on the World Wide Web". To learn about the availability of recent computerized census sources, you may want to contact the Census Bureau directly at their Customer Services division office. However, a good way to learn about the variety of census publications and other data products from the census would be to explore what is available at your college or university library.

The census data that is included on your diskettes with this book were compiled by the author from computerized samples of respondents of the 1950 to 1990 censuses on the one percent (for 1950 to 1980) and five percent (for 1990) Public Use Micro-data Samples (PUMS) available on computer tapes from the Census Bureau. While prepared from "samples" of census respondents, the numbers on your data sets are statistically weighted up so that they actually represent the total U.S. populations in each of the census years. These datasets on your diskette were prepared to be accessed with the StudentChip software, and to be consistent with the key concepts and exercises in this book. More detail on other items available on the census can generally be found in the Appendix of any major census publication. In addition, the following represent good sources of information for more specialized uses of the 1990 census for research and planning:

Richard Barrett. Using the 1990 U.S. Census for Research. Thousand Oaks,
   CA: Sage Publications, 1994.

Dowell Myers. Analysis with Local Census Data: Portraits of Change. Boston:
   Academic Press, 1992.

You may also want to know how to get information about the U.S. population since the most recent census. While it is not possible to get the kind of detailed and comprehensive information that the census provides, smaller national surveys and other data sources do provide information that will allow you to do some updating of the census. A good reference for this current information is the *Annual Statistical Abstract of the United States*, compiled by the U.S. Census Bureau. This abstract is available at most libraries.

RESOURCES ON THE WORLD WIDE WEB –

This book is being written in the midst of an explosion of available *home page* sites on the World Wide Web. While the kinds of information available on the web will undoubtedly proliferate further in the next several years, we want to call your attention to a few sites that are likely to help instructors or students using this book.

SSDAN — Social Science Data Analysis Network (http://www.psc.lsa.umich.edu/SSDAN/)

The SSDAN site was established by the author of this book with funding from the Department of Education FIPSE and the National Science Foundation Undergraduate Faculty Enhancement program. The site provides an information and data exchange mechanism for instructors interested in adopting the use of census data or other related sources in their undergraduate social science courses. This Web page will serve to update instructors on developments related to the content of this book, and other relevant resources and activities. While there is no charge, we encourage all instructors to "register" with us, on the home page, for the distribution of future messages and bulletins. (See address on next page.)

U.S. *Census Bureau* (http://www.census.gov/)

The U.S. Census Bureau has established a very useful home page that provides information about census publications, activities and the downloading of data sets. A number of colorful data displays and extraction capabilities are also available — including thematic maps of U.S. States and counties based on a variety of 1990 census characteristics.

Population Reference Bureau (http://www.prb.org/prb)

The Population Reference Bureau, Inc. is a nonprofit research organization devoted to disseminating information about population issues and topics. Its web site is an excellent source for publications, fact sheets and other materials useful in the teaching of population-related courses. Their home page enables browsers to examine and download population statistics, as well as information on current materials relevant to the classroom.

American Demographics (http://www.demographics.com)

American Demographics magazine is tailored primarily to the business audience. However, its articles are often quite useful for undergraduate population courses. Copies of articles in recent editions of the magazine are available for reading or downloading from their World Wide Web page.

Social Science Data Analysis Network

SSDAN Director William Frey

Population Studies Center
University of Michigan
1225 South University
Ann Arbor, MI 48104

http://www.psc.lsa.umich.edu/SSDAN/

Other Internet Resources

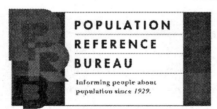

http://www.prb.org/prb

http://www.demographics.com

http://www.census.gov/

INDEX OF *Key Concepts*

The concepts used in this book's exercises are defined in the *Key Concepts* sections of the Investigation Topic chapters, in which they are featured. For easy reference, each of these concepts are listed below along with the investigation topic and page number where it is defined.

Concept Name	Topic	Page Number
Latino Groups	two	48
Marital Status	five	108
Mobility Limitations	ten	190
Occupation	two	49
Older Population	ten	190
Origin Country	three	68
Ownership	seven	139
Part-time Workers	four	90
Percent Unemployed	four	90
Percent in Labor Force	four	90
Poverty Status of a Family	eight	158
Poverty Status of a Family	eight	158
Poverty Status of a Person	eight	158
Presence of Children Under 18	seven	138
Private Schools	nine	174
Public Schools	nine	174
Race/Ethnicity	two	48
Region	one	28
Rentership	seven	139
School Aged Population	nine	174
Self-Care Limitations	ten	190
State	one	28
Unemployment Rate	four	90
Unmarried Partner	five	108
Work Disability	ten	190
Year	one	28
Year-round Full-time Worker	four	90

GUIDE TO DATASETS

The following pages provide a comprehensive guide to the datasets that are needed for the exercises in this book. They are divided into two parts. The first part lists the datasets which look at trends over the period 1950-90, called CENTREND, and appear in a separate folder (or subdirectory) on your disk. The second section pertains to datasets created from the 1990 census only and appear in the folder (or subdirectory) called CEN1990 on your disk.

When you are doing an exercise that pertains to an over-time analysis, that dataset can be found on the CENTREND list. When you are doing an exercise that pertains only to 1990, that dataset will appear on the CEN1990 list. Datasets for 1990 will also include a "9" in the name. (Names which have suffixes: "9-35" or "9-W" designate 1990 datasets specific to a specific group; 9-35 pertains to the 35-45 age group, and 9-W pertains to women, etc.)

These lists can be useful to you because they indicate the variable names that appear in each dataset, and the names of the categories associated with each variable. Since they are in alphabetical order, they appear in the same order that you will find when you scroll through the datasets on your diskette.

Shown below is an explanation of the information which appears on each dataset listing.

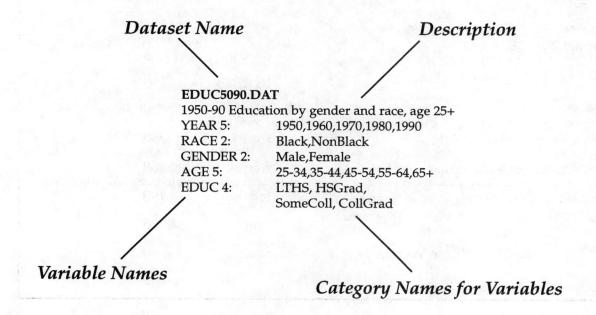

Dataset Name   *Description*

EDUC5090.DAT
1950-90 Education by gender and race, age 25+
YEAR 5:   1950,1960,1970,1980,1990
RACE 2:   Black,NonBlack
GENDER 2:   Male,Female
AGE 5:   25-34,35-44,45-54,55-64,65+
EDUC 4:   LTHS, HSGrad,
                 SomeColl, CollGrad

Variable Names

Category Names for Variables

CENTREND Datasets

BORN5090.DAT
1950-90 Children ever born for black and non-black women age 15+

YEAR 5:	1950,1960,1970,1980,1990
RACE 2:	NonBlack,Black
AGE 6:	15-24,25-34,35-44,45-54,55-64,65+
CHILD 8:	1,2,3,4,5,6+,NoneEM,NoneNM
EDUC 4:	LTHS,HSGrad, SomeColl,CollGrad

EDOC5090.DAT
1950-90 Education by occupation, race, and gender, age 16+

YEAR 5:	1950,1960,1970,1980,1990
RACE 2:	NonBlack,Black
GENDER 2:	Male,Female
WKAGE 6:	16-24,25-34,35-44,45-54,55-64,65+
OCCUP 4:	TopWC,OtrWC,Service,BC
EDUC 4:	LTHS,HSGrad, SomeColl,CollGrad

EDUC5090.DAT
1950-90 Education by gender, and race, age 25+

YEAR 5:	1950,1960,1970,1980,1990
RACE 2:	NonBlack,Black
GENDER 2:	Male,Female
AGE 5:	25-34,35-44,45-54,55-64,65+
EDUC 4:	LTHS, HSGrad, SomeColl, CollGrad

ELD5090.DAT
1950-1990 Elderly by gender, marital status, age groups

AGEELDR 3:	65-74,75-84,85+
YEAR 5:	1950,1960,1970,1980,1990
GENDER 2:	Male,Female
MARITAL 5:	CurMrrd,Widowed,Divorced, Seprated,NevMrrd

EMP5090.DAT
1950-90, Employment Status by race, marital status, and gender, age 16+

YEAR 5:	1950,1960,1970,1980,1990
RACE 2:	NonBlack,Black
EMP 3:	Empd,Unempd, NILF
WKAGE 6:	16-24,25-34,35-44,45-54,55-64,65+
MARITAL 5:	CurMrrd,Widowed,Divorced, Seprated,NevMrrd
GENDER 2:	Male,Female

FPOV7090.DAT
1970-90 Families by race, age, marital status, and poverty status, age 15+

YEAR 3:	1970,1980,1990
RACE 2:	Black,NonBlack
AGE 6:	15-24,25-34,35-44,45-54,55-64,65+
FAMTYPE 3:	MrrdCpl,MaleFam,FemlFam
POV3:	Poverty, NearPoor, Other

HH5090.DAT
1950-90 Households by household type, household size, race, age 15+

YEAR 5:	1950,1960,1970,1980,1990
HHTYPE 5:	MrrdCpl,MaleFam,FemlFam, MaleNonf,FemlNonf
HHSIZE 5:	1,2,3,4,5+
RACE 2:	NonBlack,Black
AGE 6:	15-24,25-34,35-44,45-54,55-64,65+

MARR5090.DAT
1950-90 Marital status by race, and gender, age 15+

YEAR 5:	1950,1960,1970,1980,1990
RACE 2:	NonBlack,Black
GENDER 2:	Male,Female
AGE 6:	15-24,25-34,35-44,45-54,55-64,65+
MARITAL 5:	CurMrrd,Widowed,Divorced, Seprated,NevMrrd

POP5090.DAT
1950-90 U.S. Population by age and gender for blacks and non-blacks

YEAR 5:	1950,1960,1970,1980,1990
RACE 2:	NonBlack,Black
GENDER 2:	Male,Female
AGE 8:	0-4,5-14,15-24,25-34,35-44, 45-54,55-64,65+

POPSTRUC.DAT
1930-2000 U.S. Population (in 1,000s) by age groups and gender

AGE 8:	0-4,5-14,15-24,25-34,35-44, 45-54,55-64,65+
GENDER 2:	Male,Female
YEAR 8:	1930,1940,1950,1960,1970,1980, 1990,2000

PPOV7090.DAT

1970-1990 U.S. Population by age groups, race, age, gender, and poverty status

YEAR 3:   1970,1980,1990
RACE 2:   NonBlack,Black
AGEALL 8:   0-5,6-17,18-24,25-34,35-44,45-54, 55-64,65+
GENDER 2:   Male,Female
POV 2:   Poverty,NonPov

CEN1990 Datasets

ASNUSA9.DAT

1990 Asian Population by Asian group, age, gender, and immigration status

ASIAN 7:   Chinese,Japanese,Filipino, Korean,Indian,Vietname,Other
IMMIG 4:   Native,FB<1970,FB70-79,FB80-90
GENDER: Male,Female
AGE 8:   0-4,5-14,15-24,25-34,35-44,45-54, 55-64,65+

BORN9.DAT

Children ever born by education, 1990 women age 15+

RACELAT 5:   NLWhite,Black,Latino, Asian,NLOther
EDUC 4:   LTHS,HSGrad, SomeColl,CollGrad
CHILD 8:   1,2,3,4,5,6+,NoneEM,NoneNM
AGE 6:   15-24,25-34,35-44,45-54,55-64,65+

CHLDPOV9.DAT

1990 Children by race/ethnicity, poverty, household type, family size, and immigration status

AGEKIDS 2:   0-5,6-17
RACELAT 6:   NLWhite,Black,Latino,Asian, AmIndian,NLOther
POV 4:   Poverty, NearPoor, Middle, Comf
FAMTYPE 3:   MrrdCpl,MaleFam,FemlFam
FAMSIZE 5:   2,3,4,5,6+
IMM 2:   Native, Foreign

CHLDSCH9.DAT

1990 School-age children by school attendance, poverty, English proficiency, race/ethnicity, immigration status, and gender

SCHOOL 3:   Public,Private,NotEnrld
POV 4:   Poverty,NearPoor,Middle,Comf
ENGSPKG 5:   EngOnly,Verywell,Well, Notwell,Notatall
RACELAT 6:   NLWhite,Black,Latino,Asian, AmIndian,NLOther
IMM 2:   Native, Foreign
GENDER 2:   Male, Female

CHLGRAD9.DAT

1990 Education by race/ethnicity, poverty, gender and metro, age 18-24

EDUC 4:   LTHS,HSGrad, SomeColl,CollGrad
RACELAT 6:   NLWhite,Black,Latino,Asian, AmIndian,NLOther
GENDER 2:   Male, Female
POV 2:   Poverty,NonPov
GEO 3:   City,Suburb,NonMetro
AGEHS 7:   18,19,20,21,22,23,24

COHAB9-M.DAT

1990 Men cohabitors not currently married, age 15+

AGE 6:   15-24,25-34,35-44,45-54, 55-64,65+
MARSTAT 4:   NevMrrd,Divorced, Seprated,Widowed
EDUC 4:   LTHS,HSGrad, SomeColl,CollGrad
RACELAT 5:   NLWhite,Black,Latino, Asian,NLOther

COHAB9-W.DAT
1990 Women cohabitors not currently married, age 15+
AGE 6:   15-24,25-34,35-44,45-54,55-64,65+
MARSTAT 4: NevMrrd,Divorced, Seprated,Widowed
EDUC 4:   LTHS,HSGrad, SomeColl,CollGrad
RACELAT 5: NLWhite,Black,Latino, Asian,NLOther

DOCTORS9.DAT
1990 Physicians: full time, year round workers age 25-64 by race, gender and earnings
RACELAT 6: NLWhite,Black,Asian,Latino, AmIndian,NLOther
GENDER 2:   Male,Female
AGE 4:   25-34,35-44,45-54,55-64
EARNING 8: <40K,40-55K,55-70K,70-85K, 85-100K,100-125K,125-150K,150K+

EARN9.DAT
1990 Full time, year round workers, age 16+
RACELAT 6: NLWhite,Black,Latino,Asian, AmIndian,NLOther
GENDER 2:   Male,Female
WKAGE 6:   16-24,25-34,35-44,45-54,55-64,65+
EARNING 5: <15K,15-25K,25-35K,35-50K,50K+

EARNAS9A.DAT
1990 Asian groups: full time, year round workers age 25-34
ASIAN 7:   Chinese,Japanese,Filipino, Korean,Indian,Vietname,Other
GENDER 2:   Male,Female
IMMIG 4:   Native,FB<1970,FB70-79,FB80-90
EARNING 5: <15K,15-25K,25-35K,35-50K,50K+

EARNASN9.DAT
1990 Asian groups: full time, year round workers age 16+
ASIAN 7:   Chinese, Japanese,Filipino, Korean,Indian,Vietname,Other
GENDER 2:   Male,Female
WKAGE 6:   16-24,25-34,35-44,45-54,55-64,65+
EARNING 5: <15K,15-25K,25-35K,35-50K,50K+

EARNLA9A.DAT
1990 Latino groups: full time, year round workers age 25-34
LATINO 6:   Mexican,PRican,Cuban, CAmeric,SAmeric,Other
GENDER 2:   Male,Female
IMMIG 4:   Native,FB<1970,FB70-79,FB80-90
EARNING 5: <15K,15-25K,25-35K,35-50K,50K+

EARNLAT9.DAT
1990 Latino groups, full time, year round workers age 16+
LATINO 6:   Mexican,PRican,Cuban, CAmeric,SAmeric,Other
GENDER 2:   Male,Female
WKAGE 6:   16-24,25-34,35-44,45-54,55-64,65+
EARNING 5: <15K,15-25K,25-35K,35-50K,50K+

EDOCC9.DAT
1990 Education by occupation, race, and gender for age 25+
RACELAT 5: NLWhite,Black,Latino, Asian,NLOther
GENDER 2:   Male,Female
EDUC 7:   0-9Yrs,10-12Yrs,HSGrad,SomeColl, CollGrad,Masters,PhD-Prof
OCCUP 5:   TopWC,OtrWC,Service, TopBC,OtrBC
AGE 5:   25-34,35-44,45-54,55-64,65+

EDUASN9.DAT
1990 Education for Asian groups, age 25+
ASIAN 7:   Chinese,Japanese,Filipino, Korean,Indian,Vietname,Other
AGE 5:   25-34,35-44,45-54,55-64,65+
GENDER 2:   Male,Female
EDUC 7:   0-9Yrs,10-12Yrs,HSGrad,SomeColl, CollGrad,Masters,PhD-Prof

EDUASN9A.DAT
1990 Education and immigration status for Asian groups, age 25-34
ASIAN 7:   Chinese,Japanese,Filipino, Korean,Indian,Vietname,Other
IMMIG 4:   Native,FB<1970,FB70-79,FB80-90
GENDER 2:   Male,Female
EDUC 7:   0-9Yrs,10-12Yrs,HSGrad,SomeColl, CollGrad,Masters,PhD-Prof

EDUCIMM9.DAT
1990 Education by immigration status, race/ ethnicity, gender, age 25+
EDUC 7:   0-9Yrs,10-12Yrs,HSGrad,SomeColl, CollGrad,Masters,PhD-Prof
IMM 4:   Native,FB<1970,FB70-79,FB80-90
RACELAT 6: NLWhite,Black,Latino,Asian, AmIndian,NLOther
GENDER 2:   Male,Female
AGE 5:   25-34,35-44,45-54,55-64,65+

EDULAT9.DAT

1990 Education for Latino groups, age 25+

LATINO 6:	Mexican,PRican,Cuban, CAmeric,SAmeric,Other
AGE 5:	25-34,35-44,45-54,55-64,65+
GENDER 2:	Male,Female
EDUC 7:	0-9Yrs,10-12Yrs,HSGrad,SomeColl, CollGrad,Masters,PhD-Prof

EDULAT9A.DAT

1990 Education and immigration status for Latinos groups, age 25-34

LATINO 6:	Mexican,PRican,Cuban, CAmeric,SAmeric,Other
IMMIG 4:	Native,FB<1970,FB70-79,FB80-90
GENDER 2:	Male,Female
EDUC 7:	<9Yrs,10-12Yrs,HSGrad,SomeColl CollGrad,Masters,PhD-Prof

ELDDSAB9.DAT

1990 Elderly by disability, gender, race/ethnicity, and poverty

AGEELDR 3:	65-74,75-84,85+
GENDER 2:	Male,Female
RACELAT 6:	NLWhite,Black,Latino,Asian, AmIndian,NLOther
POV 4:	Poverty, NearPoor, Middle, Comf
MOBLMT 2:	Yes,No
WRKLMT 2:	Yes,No
SELFCARE 2:	Yes,No

ELDEMP9.DAT

1990 Elderly by employment, education, poverty, and gender, age 65+

AGEELDR 3:	65-74,75-84,85+
GENDER 2:	Male,Female
EMP 4:	NILF,Unempd,EmpFull,EmpPart
EDUC 4:	LTHS,HSGrad, SomeColl,CollGrad
POV 4:	Poverty,NearPoor,Middle,Comf

ELDHH9.DAT

1990 Elderly households by age groups, gender,houshold type, and race/ethnicity

AGEELDR 3:	65-74,75-84,85+
HHTYPE 5:	MrrdCpl,MaleFam,FemlFam, MaleNonf,FemlNonf
GENDER 2:	Male,Female
RACELAT 6:	NLWhite,Black,Latino,Asian, AmIndian,NLOther

ELDPOV9.DAT

1990 Elderly by poverty, marital status, race/ethnicity, and gender, age 65+

AGEELDR 3:	65-74,75-84,85+
GENDER 2:	Male,Female
RACELAT 6:	NLWhite,Black,Asian,Latino, AmIndian,NLOther
MARITAL 5:	CurMrrd,Widowed,Divorced, Seprated,NevMrrd
POV 4:	Poverty,NearPoor,Middle,Comf

EMPASN9.DAT

1990 Employment status for Asian groups, age 16+

ASIAN 7:	Chinese,Japanese,Filipino, Korean,Indian,Vietname,Other
GENDER 2:	Male,Female
WKAGE 6:	16-24,25-34,35-44,45-54,55-64,65+
EMP 3:	Empd,Unempd,NILF

EMPASN9A.DAT

1990 Employment status by immigration status for Asian groups, age 16-34

ASIAN 7:	Chinese,Japanese,Filipino, Korean,Indian,Vietname,Other
GENDER 2:	Male,Female
WKAGE 2:	16-24,25-34
IMMIG 4:	Native,FB<1970,FB70-79,FB80-90
EMP 3:	Empd,Unempd,NILF

EMPED9.DAT

1990 Employment status by race/ethnicity, education, and gender, age 16+

RACELAT 6:	NLWhite,Black,Latino,Asian, AmIndian,NLOther
EMP 3:	Empd,Unempd,NILF
WKAGE 6:	16-24,25-34,35-44,45-54,55-64,65+
EDUC 4:	LTHS,HSGrad, SomeColl,CollGrad
GENDER 2:	Male,Female

EMPLAT9.DAT

1990 Employment status for Latino groups, age 16+

LATINO 6:	Mexican,PRican,Cuban, CAmeric,SAmeric,Other
GENDER 2:	Male,Female
WKAGE 6:	16-24,25-34,35-44,45-54,55-64,65+
EMP 3:	Empd,Unempd,NILF

EMPLAT9A.DAT

1990 Employment status by immigration status
for Latino groups, age 16-34
LATINO 6:   Mexican,PRican,Cuban,
               CAmeric,SAmeric,Other
GENDER 2:   Male,Female
WKAGE 2:   16-24,25-34
IMMIG 4:   Native,FB<1970,FB70-79,FB80-90
EMP 3:   Empd,Unempd,NILF

EMPLOY9.DAT

1990 Employment status for age 16+
RACELAT 6: NLWhite,Black,Latino,Asian,
               AmIndian,NLOther
GENDER 2:   Male,Female
WKAGE 6:   16-24,25-34,35-44,45-54,55-64,65+
EMP 3:   Empd,Unempd,NILF

ENGASN9.DAT

1990 Asian groups by immigration status,
English proficiency, gender, state, and age
IMM 4:   Native,FB<1970,FB70-79,FB80-90
ENGSPKG 5: EngOnly,Verywell,Well,
               Notwell,Notatall
ASIAN 7:   Chinese,Japanese,Filipino,
               Korean,Indian,Vietname,Other
GENDER 2:   Male,Female
AGE 3:   0-17,18-64,65+
STATECAL 2: CA,RestofUS

ENGLAT9.DAT

1990 Latino groups by immigration status,
English proficiency, gender, state, and age
IMM 4:   Native,FB<1970,FB70-79,FB80-90
ENGSPKG 5: EngOnly,Verywell,Well,
               Notwell,Notatall
LATINO 6:   Mexican,Cuban,PRican,
               CAmeric,SAmeric,Other
GENDER 2:   Male,Female
AGE 3:   0-17,18-64,65+
STATECAL 2: CA,RestofUS

FAMILY9.DAT

1990 Families by family type, gender, poverty,
and race/ethnicity, age 15+
FAMTYPE 3: MrrdCpl,MaleFam,FemlFam
RACELAT 6: NLWhite,Black,Latino,Asian,
               AmIndian,NLOther
KID 3:   None,Kids<6,KidsOtr
POV 4:   Poverty, NearPoor, Middle, Comf
AGE 6:   <25,25-34,35-44,45-54,55-64,65+

FERNTYP9.DAT

1990 Married couple families by earners age 16+
by race/ethnicity, and poverty
EARNTYP 4: 2ErnrFam,1ErnrML,
               1ErnrFML,None
RACELAT 6: NLWhite,Black,Latino,Asian,
               AmIndian,NLOther
POV 4:   Poverty, NearPoor, Middle, Comf
AGE 6:   <25,25-34,35-44,45-54,55-64,65+

FPOV9.DAT

1990 Families by family type and poverty status
RACELAT 6: NLWhite,Black,Asian,Latino,
               AmIndian,NLOther
AGE 6:   <25,25-34,35-44,45-54,55-64,65+
FAMTYPE 3: MrrdCpl,MaleFam,FemlFam
POV 2:   Poverty,NonPov

FPOVASN9.DAT

1990 Family type by poverty status for Asian
groups, all ages
ASIAN 7:   Chinese,Japanese,Filipino,
               Korean,Indian,Vietname,Other
AGE 6 :   <25,25-34,35-44,45-54,55-64,65+
FAMTYPE 3: MrrdCpl,MaleFam,FemlFam
POV 2:   Poverty,NonPov

FPOVGEO9.DAT

1990 Families by familiy type, race/ethnicity,
poverty, and metro, age 15+
RACELAT 6: NLWhite,Black,Latino,Asian,
               AmIndian,NLOther
POV 4:   Poverty, NearPoor, Middle, Comf
FAMTYPE 3: MrrdCpl,MaleFam,FemlFam
GEO 3:   City,Suburb,NonMetro
AGE 6:   <25, 25-34,35-44,45-54,55-64,65+

FPOVLAT9.DAT

1990 Family type by poverty status for Latino
groups, all ages
LATINO 6:   Mexican,PRican,Cuban,
               CAmeric,SAmeric,Other
AGE 6:   <25,25-34,35-44,45-54,55-64,65+
FAMTYPE 3: MrrdCpl,MaleFam,FemlFam
POV 2:   Poverty,NonPov

FPRASN9.DAT
1990 Work hours for Asian groups, age 16+
ASIAN 7:	Chinese,Japanese,Filipino, Korean,Indian,Vietnam,Other
GENDER 2:	Male,Female
WKAGE 6:	16-24,25-34,35-44,45-54,55-64,65+
WKHRS 4:	Full35,20-34,10-19,<10

FPRASN9A.DAT
1990 Work hours by immigration status for Asian groups, age 16-34
ASIAN 7:	Chinese,Japanese,Filipino, Korean,Indian,Vietnam,Other
GENDER 2:	Male,Female
WKAGE 2:	16-24,25-34
IMMIG 4:	Native,FB<1970,FB70-79,FB80-90
WKHRS 4:	Full35,20-34,10-19,<10

FPRLAT9.DAT
1990 Work hours for Latino groups, age 16+
LATINO 6:	Mexican,PRican,Cuban, CAmeric,SAmeric,Other
GENDER 2:	Male,Female
WKAGE 6:	16-24,25-34,35-44,45-54,55-64,65+
WKHRS 4:	Full35,20-34,10-19,<10

FPRLAT9A.DAT
1990 Work hours by immigration status for Latino groups, age 16-34
LATINO 6:	Mexican,PRican,Cuban, CAmeric,SAmeric,Other
GENDER 2:	Male,Female
WKAGE 2:	16-24,25-34
IMMIG 4:	Native,FB<1970,FB70-79,FB80-90
WKHRS 4:	Full35,20-34,10-19,<10

FPRTED9.DAT
1990 Work hours by race/ethnicity, gender, education, age 16+
RACELAT 6:	NLWhite,Black,Latino,Asian, AmIndian,NLOther
GENDER 2:	Male, Female
WKAGE 6:	16-24,25-34,35-44,45-54,55-64,65+
EDUC 4:	LTHS,HSGrad, SomeColl,CollGrad
WKHRS 4:	Full35,20-34,10-19,<10

FULLPRT9.DAT
1990 Workhours for workers, age 16+
RACELAT 6:	NLWhite,Black,Latino,Asian, AmIndian,NLOther
GENDER 2:	Male,Female
WKAGE 6:	16-24,25-34,35-44,45-54,55-64,65+
WKHRS 4:	Full35,20-34,10-19,<10

HHOLDS9.DAT
1990 Households by household type, poverty, and race/ethnicity, age 15+
AGE 6:	15-24,25-34,35-44,45-54,55-64,65+
HHTYPE 5:	MrrdCpl,MaleFam,FemlFam, MaleNonf, FemNonf
POV 4:	Poverty, NearPoor, Middle, Comf
RACELAT 6:	NLWhite,Black,Latino,Asian, AmIndian,NLOther

HOUSNG9.DAT
1990 Households by household type, housing, race/ethnicity, and homeownership, age 15+
HOUSING 5:	House,Apt2-9,Apt10+, MobHome,Other
RACELAT 6:	NLWhite,Black,Latino,Asian, AmIndian,NLOther
HHTYPE 5:	MrrdCpl,MaleFam,FemlFam, MaleNonf,FemlNonf
AGE 6:	15-24,25-34,35-44,45-54,55-64,65+
HOMEOWNR 2:	Homeownr, Renter

IMMUSA9.DAT
1990 U.S. Population by race/ethnicity, gender, poverty, immigration status, and state
STATEUSA 8:	CA,FL,IL,NJ,NY,TX,VT,RestofUS
IMM 4:	Native,FB<1970,FB70-79,FB80-90
RACELAT 6:	NLWhite,Black,Latino,Asian, AmIndian,NLOther
GENDER 2:	Male,Female
POV 4:	Poverty,NearPoor, Middle,Comf

KIDEMP9.DAT
1990 Employment status by number of children, education for currently married women, age 25-34
RACELAT 5:	NLWhite,Black,Latina, Asian,NLOther
EDUC 4:	LTHS,HSGrad, SomeColl,CollGrad
KID 3:	None,Kids<6,KidsOtr
EMP 4:	NILF,Unempd,EmpFull,EmpPart

LATUSA9.DAT
1990 Latino population by Latino group, age, gender and immigration status
LATINO 6:	Mexican,PRican,Cuban, CAmeric,SAmeric,Other
IMMIG 4:	Native,FB<1970,FB70-79,FB80-90
GENDER 2:	Male,Female
AGE 8:	0-4,5-14,15-24,25-34,35-44,45-54, 55-64,65+

LAWYERS9.DAT
1990 Lawyers: full time, year round workers, race, gender, and earnings, age 25-64
RACELAT 6: NLWhite,Black,Asian,Latino, AmIndian,NLOther
GENDER 2: Male,Female
AGE 4: 25-34,35-44,45-54,55-64
EARNING 8: <40K,40-55K,55-70K,70-85K, 85-100K,100-125K,125-150K,150K+

MARASN9.DAT
1990 Marital status for Asian groups, age 15+
ASIAN 7: Chinese,Japanese,Filipino, Korean,Indian,Vietname,Other
GENDER 2: Male,Female
AGE 6: 15-24,25-34,35-44,45-54, 55-64,65+
MARITAL 5: CurMrrd,Widowed,Divorced, Seprated,NevMrrd

MARED9.DAT
1990 Marital status by education, gender, and race/ethnicity,age 15+
AGE 6: 15-24,25-34,35-44,45-54,55-64,65+
GENDER 2: Male,Female
RACELAT 6: NLWhite,Black,Latino,Asian, AmIndian,NLOther
EDUC 4: LTHS,HSGrad, SomeColl,CollGrad
MARITAL 5: CurMrrd,Divorced,Seprated, Widowed,NevMrrd

MARITAL9.DAT
1990 Marital status for age 15+
RACELAT 6: NLWhite,Black,Asian,Latino, AmIndian,NLOther
GENDER 2: Male,Female
AGE 6: 15-24,25-34,35-44,45-54,55-64,65+
MARITAL 5: CurMrrd,Widowed,Divorced, Seprated,NevMrrd

MARLAT9.DAT
1990 Marital status for Latino groups, age 15+
LATINO 6: Mexican,PRican,Cuban, CAmeric,SAmeric,Other
GENDER 2: Male,Female
AGE 6: 15-24,25-34,35-44,45-54,55-64,65+
MARITAL 5: CurMrrd,Widowed,Divorced, Seprated,NevMrrd

MREMPF9.DAT
1990 Females by marital status, employment, and poverty , race/ethnicity, age 16+
WKAGE 6: 16-24,25-34,35-44,45-54,55-64,65+
MARITAL 5: CurMrrd,Divorced,Seprated, Widowed,NevMrrd
EMP 4: NILF, Unempd,EmpFull, EmpPart
POV 2: Poverty, NonPov
RACELAT 6: NLWhite,Black,Latina,Asian, AmIndian,NLOther

MRR9-YM.DAT
1990 Marital/cohab status for young men age 23-28
YAGE 6: 23,24,25,26,27,28
MARSTUS 4: NevMrrd,LivingTg, CurMrrd,DivSepWid
EDUC 4: LTHS,HSGrad, SomeColl,CollGrad
RACELAT 5: NLWhite,Black,Latino, Asian,NLOther

MRR9-YW.DAT
1990 Marital/cohab status for young women age 23-28
YAGE 6: 23,24,25,26,27,28
MARSTUS 4: NevMrrd,LivingTg, CurMrrd,DivSepWid
EDUC 4: LTHS,HSGrad, SomeColl,CollGrad
RACELAT 5: NLWhite,Black,Latina, Asian,NLOther

MRREMP9.DAT
1990 Marital/cohab status by Employment status for women age 25-34
RACELAT 5: NLWhite,Black,Latina, Asian,NLOther
EDUC 4: LTHS,HSGrad, SomeColl,CollGrad
MARSTAT 6: NevMrrd,LivingTg,CurMrrd, Divorced,Seprated,Widowed
EMP 4: NILF,Unempd,EmpFull,EmpPart

OCCASN9.DAT
1990 Occupations for Asian groups, age 16+
ASIAN 7: Chinese,Japanese,Filipino, Korean,Indian,Vietname,Other
GENDER 2: Male,Female
WKAGE 6: 16-24,25-34,35-44,45-54,55-64,65+
OCCUP 6: TopWC,OtrWC,Service, TopBC,OtrBC,Farm

OCCASN9A.DAT
1990 Occupation and immigration status for
Asian groups, age 25-34
ASIAN 7:   Chinese,Japanese,Filipino,
            Korean,Indian,Vietname,Other
GENDER 2: Male,Female
IMMIG 4:   Native,FB<1970,FB70-79,FB80-90
OCCUP 6:   TopWC,OtrWC,Service,
            TopBC,OtrBC,Farm

OCCLAT9.DAT
1990 Occupations for Latino groups, age 16+
LATINO 6: Mexican,PRican,Cuban,
            CAmeric,SAmeric,Other
GENDER 2: Male,Female
WKAGE 6:   16-24,25-34,35-44,45-54,55-64,65+
OCCUP 6:   TopWC,OtrWC,Service,
            TopBC,OtrBC,Farm

OCCLAT9A.DAT
1990 Occupation and immigration status for
Latino groups, age 25-34
LATINO 6: Mexican,PRican,Cuban,
            CAmeric,SAmeric,Other
GENDER 2: Male,Female
IMMIG 4:   Native,FB<1970,FB70-79,FB80-90
OCCUP 6:   TopWC,OtrWC,Service,
            TopBC,OtrBC,Farm

OCCUPTN9.DAT
1990 Occupations, age 16+
RACELAT 6: NLWhite,Black,Asian,Latino,
            AmIndian,NLOther
GENDER 2: Male,Female
WKAGE 6:   16-24,25-34,35-44,45-54,55-64,65+
OCCUP 6:   TopWC,OtrWC,Service,
            TopBC,OtrBC,Farm

OCIM9-25.DAT
1990 Full-time year round workers by occupa-
tion, gender, immigration status, race/ethnicity,
and education age 25-34
OCCUP 5:   TopWC,OtrWC,Service,
            TopBC,OtrBC
GENDER 2: Male, Female
IMM 4:   Native,FB<1970,FB70-79,FB80-90
RACELAT 6: NLWhite,Black,Latino,Asian,
            AmIndian,NLOther
EDUC 7:   0-9Yrs,10-12Yrs,HSGrad,SomeColl
            CollGrad,Masters,PhD-Prof

OCIM9-35.DAT
1990 Full-time year round workers by occupa-
tion, gender, immigration status, race/ethnicity,
and education age 35-44
OCCUP 5:   TopWC,OtrWC,Service
            TopBC,OtrBC
GENDER 2: Male, Female
IMM 4:   Native,FB<1970,FB70-79,FB80-90
RACELAT 6: NLWhite,Black,Latino,Asian,
            AmIndian,NLOther
EDUC 7:   0-9Yrs,10-12Yrs,HSGrad,SomeColl,
            CollGrad,Masters,PhD-Prof

POPCA9.DAT
1990 California population by age, gender, race
and immigration status
RACELAT 6: NLWhite,Black,Asian,Latino,
            AmIndian,NLOther
IMMIG 4:   Native,FB<1970,FB70-79,FB80-90
GENDER 2: Male,Female
AGE 8:   0-4,5-14,15-24,25-34,35-44,45-54,
            55-64,65+

POPGA9.DAT
1990 Georgia population by age, gender, race
and immigration status
RACELAT 6: NLWhite,Black,Asian,Latino,
            AmIndian,NLOther
IMMIG 4:   Native,FB<1970,FB70-79,FB80-90
GENDER 2: Male,Female
AGE 8:   0-4,5-14,15-24,25-34,35-44,45-54,
            55-64,65+

POPGEO9.DAT
1990 U.S. Population by age groups, metro,
regions, and race/ethnicity
AGE 8:   0-4,5-14,15-24,25-34,35-44,45-54,
            55-64,65+
GEO 3:   City,Suburb,NonMetro
REGION 4:   Nrtheast,Midwest,South,West
RACELAT 6: NLWhite,Black,Latino,Asian,
            AmIndian,NLOther

POPLA9.DAT
1990 L.A. County population, age, gender, race
and immigration status
RACELAT 6: NLWhite,Black,Asian,Latino,
            AmIndian,NLOther
IMMIG 4:   Native,FB<1970,FB70-79,FB80-90
GENDER 2: Male,Female
AGE 8:   0-4,5-14,15-24,25-34,35-44,45-54,
            55-64,65+

POPMI9.DAT
1990 Michigan population by age, gender, race and immigration status

RACELAT 6:	NLWhite,Black,Asian,Latino, AmIndian,NLOther
IMMIG 4:	Native,FB<1970,FB70-79,FB80-90
GENDER 2:	Male,Female
AGE 8:	0-4,5-14,15-24,25-34,35-44,45-54, 55-64,65+

POPPROJ9.DAT
Projected population by race/ethnicity, state, and age

AGE 3:	0-17,18-64,65+
RACELT 5:	NLWhite,Black,Latino,Asian, AmIndian
PROJYEAR 6:	1995,2000,2005,2010,2015,2020
STATEUSA 8:	CA,FL,IL,NJ,NY,TX,VT,RestofUS

POPUSA9.DAT
1990 US population by age, gender, race and immigration status

RACELAT 6:	NLWhite,Black,Asian,Latino, AmIndian,NLOther
IMMIG 4:	Native,FB<1970,FB70-79,FB80-90
GENDER 2:	Male,Female
AGE 8:	0-4,5-14,15-24,25-34,35-44,45-54, 55-64,65+

PPOVEDU9.DAT
1990 U.S. Population by race/ethnicity, poverty, gender, and education, age 25+

RACELAT 6:	NLWhite,Black,Latino,Asian, AmIndian,NLOther
POV 4:	Poverty, NearPoor, Middle, Comf
GENDER 2:	Male,Female
AGE 5:	25-34,35-44,45-54,55-64,65+
EDUC 7:	0-9Yrs,10-12Yrs,HSGrad,SomeColl, CollGrad,Masters,PhD-Prof

PPOVGEO9.DAT
1990 U.S. Population by age groups, race/ ethnicity, poverty, metro, and gender

RACELAT 6:	NLWhite,Black,Latino,Asian, AmIndian,NLOther
POV 4:	Poverty, NearPoor, Middle, Comf
GENDER 2:	Male,Female
AGEALL 8:	0-5,6-17,18-24,25-34,35-44,45-54, 55-64,65+
GEO 3:	City,Suburb,NonMetro

SPAGE9YM.DAT
1990 Spouses: married men age 25 by education and wife's age 15-29+

HRACELAT 5:	NLWhite,Black,Latino, Asian,NLOther
HEDUC 4:	LTHS,HSGrad, SomeColl,CollGrad
WAGE 8:	<23,23,24,25,26,27,28,29+

SPAGE9YW.DAT
1990 Spouses: married women age 25 by education and husband's age 15-29+

WRACELAT 5:	NLWhite,Black,Latina, Asian,NLOther
WEDUC 4:	LTHS,HSGrad, SomeColl,CollGrad
HAGE 8:	<23,23,24,25,26,27,28,29+

SPAGER9.DAT
1990 Spouses: husband's and wife's age 15-45+ and husband's and wife's race

HRACELAT 5:	NLWhite,Black,Latino, Asian,NLOther
WRACELAT 5:	NLWhite,Black,Latina, Asian,NLOther
HAGE 4:	15-24,25-34,35-44,45+
WAGE 4:	15-24,25-34,35-44,45+

SPED9-M.DAT
1990 Spouses: married men age 15+, by race and education, by wife's education

HAGE 6:	15-24,25-34,35-44,45-54,55-64,65+
HRACELAT 5:	NLWhite,Black,Latino, Asian,NLOther
HEDUC 4:	LTHS,HSGrad, SomeColl,CollGrad
WEDUC 4:	LTHS,HSGrad, SomeColl,CollGrad

SPED9-W.DAT
1990 Spouses: married women age 15+, by race and education, by husband's education

WAGE 6:	15-24,25-34,35-44,45-54,55-64,65+
WRACELAT 5:	NLWhite,Black,Latina, Asian,NLOther
WEDUC 4:	LTHS,HSGrad, SomeColl,CollGrad
HEDUC 4:	LTHS,HSGrad, SomeColl,CollGrad

SPEDR9.DAT
1990 Spouses: husband and wife's education and race
HRACELAT 5: NLWhite,Black,Latino, Asian,NLOther
WRACELAT 5: NLWhite,Black,Latina, Asian,NLOther
HEDUC 4:   LTHS,HSGrad, SomeColl,CollGrad
WEDUC 4:   LTHS,HSGrad, SomeColl,CollGrad

SPRAC9-M.DAT
1990 Spouses: married men age 15+, by education and race, by wife's race
HAGE 5:   15-24,25-34,35-44,45-54,55+
HEDUC 4:   LTHS,HSGrad, SomeColl,CollGrad
HRACELAT 5: NLWhite,Black,Latino, Asian,NLOther
WRACELAT 5: NLWhite,Black,Latina, Asian,NLOther

SPRAC9-W.DAT
1990 Spouses: married women age 15+, by education and race, by husband's race
WAGE 5:   15-24,25-34,35-44,45-54,55+
WEDUC 4:   LTHS,HSGrad, SomeColl,CollGrad
WRACELAT 5: NLWhite,Black,Latina, Asian,NLOther
HRACELAT 5: NLWhite,Black,Latino, Asian,NLOther

WKAS9-25.DAT
1990 Full-time year round Asian workers by earnings, immigration status, education, occupation , and gender, age 25-34
EARN 5:   <15K,15-25K,25-35K,35-50K,50K+
IMM 4:   Native,FB<1970,FB70-79,FB80-90
EDUC 4:   LTHS,HSGrad, SomeColl,CollGrad
OCCUP 5:   TopWC,OtrWC,Service, TopBC,OtrBC
GENDER 2:   Male,Female

WKAS9-35.DAT
1990 Full-time year round Asian workers by earnings, immigration status, education, occupation , and gender, age 35-44
EARN 5:   <15K,15-25K,25-35K,35-50K,50K+
IMM 4:   Native,FB<1970,FB70-79,FB80-90
EDUC 4:   LTHS,HSGrad, SomeColl,CollGrad
OCCUP 5:   TopWC,OtrWC,Service, TopBC,OtrBC

WKIM9-25.DAT
1990 Full-time year round workers by earnings, immigration status, education, occupation, and gender, age 25-34
EARN 5:   <15K,15-25K,25-35K,35-50K,50K+
IMM 4:   Native,FB<1970,FB70-79,FB80-90
EDUC 4:   LTHS,HSGrad, SomeColl,CollGrad
OCCUP 5:   TopWC,OtrWC,Service, TopBC,OtrBC
GENDER 2:   Male,Female

WKIM9-35.DAT
1990 Full-time year round workers by earnings, immigration status, education , occupation, and gender, age 35-44
EARN 5:   <15K,15-25K,25-35K,35-50K,50K+
IMM 4:   Native,FB<1970,FB70-79,FB80-90
EDUC 4:   LTHS,HSGrad, SomeColl,CollGrad
OCCUP 5:   TopWC,OtrWC,Service, TopBC,OtrBC
GENDER 2:   Male,Female

WKLT9-25.DAT
1990 Full-time year round Latino workers by earnings, immigration status, education, occupation , and gender, age 25-34
EARN 5:   <15K,15-25K,25-35K,35-50K,50K+
IMM 4:   Native,FB<1970,FB70-79,FB80-90
EDUC 4:   LTHS,HSGrad, SomeColl,CollGrad
OCCUP 5:   TopWC,OtrWC,Service, TopBC,OtrBC
GENDER 2:   Male,Female

WKLT9-35.DAT

1990 Full-time year round Latino workers by earnings, immigration status, education, occupation , and gender, age 35-44

EARN 5:	<15K,15-25K,25-35K,35-50K,50K+
IMM 4:	Native,FB<1970,FB70-79,FB80-90
EDUC 4:	LTHS,HSGrad, SomeColl,CollGrad
OCCUP 5:	TopWC,OtrWC,Service, TopBC,OtrBC
GENDER 2:	Male,Female

WORK9-25.DAT

1990 Full time, year round workers by education, occupation, and earnings, age 25-34

RACELAT 3:	NLWhite,Black,AllOther
GENDER 2:	Male,Female
EDUC 4:	LTHS,HSGrad, SomeColl,CollGrad
OCCUP 4:	TopWC,OtrWC,Service,BC
EARNING 5:	<15K,15-25K,25-35K,35-50K,50K+

WORK9-35.DAT

1990 Full time, year round workers by education, occupation, and earnings, age 35-44

RACELAT 3:	NLWhite,Black,AllOther
GENDER 2:	Male,Female
EDUC 4:	LTHS,HSGrad, SomeColl,CollGrad
OCCUP 4:	TopWC,OtrWC,Service,BC
EARNING 5:	<15K,15-25K,25-35K,35-50K,50K+

WORK9-45.DAT

1990 Full time, year round workers by education, occupation, and earnings, age 45-54

RACELAT 3:	NLWhite,Black,AllOther
GENDER 2:	Male,Female
EDUC 4:	LTHS,HSGrad, SomeColl,CollGrad
OCCUP 4:	TopWC,OtrWC,Service,BC
EARNING 5:	<15K,15-25K,25-35K,35-50K,50K+